CONCENTRATION AND DRYING OF FOODS

The Proceedings of the Kellogg Foundation Second International Food Research Symposium held at University College, Cork, Republic of Ireland, 16–18 September 1985

CONCENTRATION AND DRYING OF FOODS

Edited by

Diarmuid MacCarthy
University College, Cork, Republic of Ireland

ELSEVIER APPLIED SCIENCE PUBLISHERS
LONDON and NEW YORK

ELSEVIER APPLIED SCIENCE PUBLISHERS LTD
Crown House, Linton Road, Barking, Essex IG11 8JU, England

Sole Distributor in the USA and Canada
ELSEVIER SCIENCE PUBLISHING CO., INC.
52 Vanderbilt Avenue, New York, NY 10017, USA

WITH 54 TABLES AND 97 ILLUSTRATIONS

British Library Cataloguing in Publication Data

Concentration and drying of foods.
1. Food – Drying
I. MacCarthy, Diarmuid
664'.0284 TP371.5

ISBN 0-85334-442-6

Library of Congress CIP data applied for

The selection and presentation of material and the opinions expressed in this publication are the sole responsibility of the authors concerned.

Printed in Great Britain at the University Press, Cambridge

Preface

Since earliest times the concentration and drying of foods has enabled man to preserve surplus food for use in times of scarcity. What our early ancestors learned by trial and error has developed into a world-wide industry, providing a high quality nutritious and convenient food supply for urban and rural dweller alike.

Research continues unabated into all facets of food concentration and drying, from the application of mathematical modelling to drier design, to elucidating the mechanism of chemical changes such as Maillard browning which damage food during drying.

Because of the importance of this subject to the world's food supply in general, and to the Irish food industry in particular, an international panel of speakers was assembled at University College, Cork, for the Kellogg Foundation Second International Food Research Symposium, to provide up-to-date critical reviews of recent developments in this area. The results of their deliberations are presented in these proceedings.

Our grateful thanks are due to the W. K. Kellogg Foundation whose generous sponsorship made the symposium possible, to the authors, and to the many people who contributed so greatly to the organisation of the symposium and the preparation of the proceedings, in particular to R. O'Riordan, C. Butler, P. M. Kelly and P. A. Morrissey.

DIARMUID MACCARTHY
University College, Cork

Contents

List of Contributors

F. W. BAKKER-ARKEMA
Agricultural Engineering Department, Michigan State University, East Lansing, Michigan 48824, USA.

C. VAN DEN BERG
Department of Food Science, Agricultural University, De Dreijen 12, 6703 BC Wageningen, The Netherlands.

T. C. DUANE
College of Catering, Cathal Bruga Street, Dublin 1, Ireland.

H. F. ERBERSDOBLER
Institut für Humanernährung and Lebensmittelkunde der Universität Kiel, Düsternbrooker Weg 17–19, D-2300 Kiel 1, Federal Republic of Germany.

P. A. GIBBS
Leatherhead Food Research Association, Randalls Road, Leatherhead, Surrey KT22 7RY, UK.

G. HAHN
Dairy Department, GEA-Wiegand GmbH, Einsteinstrasse 9–15, D-7505 Ettlingen, Federal Republic of Germany.

S. D. HOLDSWORTH
Campden Food Preservation Research Association, Chipping Campden, Gloucestershire GL55 6LD, UK.

M. KAREL

Department of Applied Biological Sciences, Massachusetts Institute of Technology, Cambridge, Massachusetts 02139, USA.

H. G. KESSLER

Institut für Milchwissenschaft und Lebensmittelverfahrenstechnik, Technische Universität München, D-8050 Freising-Weihenstephan, Federal Republic of Germany.

M. E. KNIPSCHILDT

APV Anhydro A/S, 8 Østmarken, DK-2860 Søborg-Copenhagen, Denmark.

J. MCKEON

Industrial Development Authority, Lansdowne House, Dublin 4, Ireland.

H. E. NURSTEN

Department of Food Science, University of Reading, PO Box 226, Reading, Berks RG6 2AP, UK.

B. OSTERGAARD

Pasilac A/S, DK-8600 Silkeborg, Denmark.

J. PISECKY

A/S Niro Atomizer, 305 Gladsaxevej, DK-2860 Søborg-Copenhagen, Denmark.

E. C. SYNNOTT

Department of Dairy and Food Engineering, University College, Cork, Ireland.

MARKET TRENDS IN DEHYDRATED FOODS

JOHN MCKEON
Industrial Development Authority, Dublin

ABSTRACT

The food market is facing enormous changes and challenges internationally. Massive food shortages in parts of the developing world coexist with huge stockpiles in the developed world. In the developed world the rate of product change is dramatic, as changing lifestyles demand better quality, healthier and more convenient foods for an increasingly ageing population living generally in smaller family units.

Frozen, chilled and fresh foods are replacing many traditionally preserved products. The retail sector is becoming more concentrated and is challenging all but the leading brands with generic and own-label products. Packaging, presentation and merchandising are all changing rapidly.

The dehydrated food sector, which world-wide probably exceeds $20 billion annually, does not appear to be making any major changes to meet the requirements of the modern consumer, and it appears to be losing market share particularly in the developed world. Such an adverse trend if sustained has implications for a large number of researchers, machine suppliers, food manufacturers and consumers. Industrialists in Ireland have a direct interest in the market trend for dried foods particularly in such areas as dairy products, soups and to a lesser degree vegetables.

This paper examines the overall demand trends emerging for dried foods, and the trends for some expanding niches in the market. It attempts to identify the consumers' perceptions of the quality of dried foods and it highlights some of the implications of these trends for the sector and for the Irish dried food industry.

GLOBAL TRENDS

Dehydration is one of the oldest forms of preserving food, and has been used for generations for preserving vegetables, fruits, coffee, cereals and fish. In more recent times the development of modern and novel drying techniques such as freeze-drying helped develop demand for a range of higher quality products, and fueled a rapid expansion in the demand for dried foods over the last two decades. The dramatic growth of the recent past has now slowed considerably and the demand outlook for dried foods in the developed world is not encouraging for the immediate future.

The dried food market is not well defined, and the quality of data and their availability are poor for many market segments. It is nonetheless possible to identify a number of reasons for the gloomy market prospects in the developed world.

Perhaps the more important of these reasons is the changing lifestyle of the modern consumer. This new lifestyle with its emphasis on high quality and healthy foods, properties which the consumer associates with freshness, has resulted in a static or declining demand for many dried foods in the developed world markets. In addition, the increased level of freezer ownership has allowed for increased use of frozen and chilled products at the expense of dried foods in many cases. Dried foods are promoted as convenient, which undoubtedly they are, but convenience alone is not sufficient for the modern housewife; quality and freshness are also regarded as cruical attributes in food products nowadays.

In the developed world, dried foods are increasingly purchased by lower income groups, and a greater share of product is moving through the less discerning catering and institutional trade rather than the retail trade. These factors influence perceptions, and they tend to reduce the return available to manufacturers and therefore their ability to re-invest and to develop and promote new products.

Frost and Sullivan (ref. 1) in a major review of trends in the European market for dried foods conclude that the outlook is disappointing and they forecast a modest growth of the order of 2% p.a. for the overall sector to 1987.

Mintel (ref.2) are more pessimistic than this, and conclude that large promotional back-up will allow the dried food market to become static around its present volume before tailing off into a long-term decline.

Euromonitor (ref.3) commenting on the UK market conclude that the market for dried foods is "on the whole static if not declining".

Basebook data (ref. 4) from the US show an annual decline of over 4.0% in nonfat dried milk, 1.8% in dry whole milk, 1.6% in dried vegetables and 0.1% in dried fruits, in the US market.

The predominant trend apparent in the major dried food markets in the developed world is of static or declining markets. Dried foods generally do best in a market niche where fresh or frozen competitive products do not exist e.g. breakfast cereals, instant coffee.

In the developing world, the information available suggests that the markets for dried foods are expanding. This expansion is notable for such products as dried milks and baby foods. Problems have been identified with dried foods in warm and humid climates where spoilage can occur unless sophisticated packaging is used. These markets, while offering enormous potential, require enormous expenditure to ensure product quality and timely delivery to consumers.

TRENDS IN THE EUROPEAN MARKET

The European market for dried foods was estimated at $7 billion in 1982 (ref. 1). The market developed rapidly during the '60's and '70's as higher quality convenient foods were added to the traditional range of dried foods such as coffee, milk and vegetables. New drying techniques helped develop this expanding market.

In recent years, the spread of home freezer ownership and the provision of a range of frozen and chilled products which are convenient and of high quality have resulted in a loss of market share by a wide range of dried food products. Frost and Sullivan's prediction (ref. 1) is that the market will grow modestly to $7.8 billion in 1987, (Table 1).

TABLE 1

Demand for all Dehydrated Foods - Europe 1982 - 1987 ($ million)

Product Category	1982	1987	% p.a. Change
Dehydrated & Convenience Foods	1896	2140	+ 2.8%
Dry Beverages	4015	4456	+ 1.9%
Dehydrated Sauces and Mixes	1167	1266	+ 1.5%
Total	7078	7862	+ 2.1%

Within these overall growth trends there are marked differences in the growth rates expected in different product categories and in different countries.

Among the main dehydrated product categories, meat substitutes, for example, are expected to grow at about 10% p.a. while the demand for products like custards is expected to decline by almost 5% p.a. (Table 2)

TABLE 2

European Market for Dehydrated Foods

Product	% Total Market	% Growth p.a.
Coffee	46	2.0
Soup	13	2.4
Dried Milk	8	.5
Dessert Mixes/Toppings	7	.6
Gravy Sauces	7	2.0
Dried Potato	6	2.7
Complete Meals	4	4.9
Meat Substitutes	0.3	9.8
Custards	0.2	- 4.4

Across countries, substantial variations are also expected to occur. (Tables 3 and 4). The UK and West Germany are the main markets for dried food and both are showing growth rates at or above the European average. On the other hand the market in the Netherlands is almost static.

TABLE 3

Per Capita Consumption of Dehydrated Foods (kg. per annum)

UK	5.5
West Germany	4.8
Netherlands	4.7
Belgium	4.0
France	3.3
Italy	1.3

TABLE 4

Consumption Trends : Dehydrated Foods

Country	% of Total Market	% Growth p.a.
UK	29	2.6
West Germany	39	2.1
France	14	1.8
Italy	10	2.2
Netherlands	5	0.8
Belgium	3	1.8
	100	2.1

The expectations for different products vary greatly among countries. Soup consumption is expanding by almost 5% p.a. in the UK but is almost static in the Netherlands. Dessert mixes are expanding by 4.0% p.a. in the UK but are declining by almost 3% p.a. in West Germany.

A further significant feature of the European market is the degree of retail versus non-retail sales in the sector.

In 1982, the retail sector absorbed 60% of the total volume of dried foods but it provided 72% of the total market revenue. (Table 5). The catering, institutional and food processing trades absorb an increasing share of product especially such items as vegetables, potatoes, soups and dried milk. Without these products the retail sector accounts for 75% of volume and almost 90% of value of dried food sales.

TABLE 5
Retail and Non-Retail Market Shares - European Dried Foods

	Retail	Non Retail
Volume of Trade	60%	40%
Value of Trade	72%	28%

Dried foods appear to be most popular with lower income families. The British Market Research Bureau found that working class households (C, D income groups) used dried food products almost twice as often as the higher income groups (A, B).

The packaging of dried foods is mostly in plastic and cartonboard, and is generally not very expensive.

TABLE 6
Packaging Materials - Dehydrated Food

Type of Package	Market Share (%)
Cartonboard outer and Bag	26
Sachet	22
Film	7
Plastic	45
	100

With some variations among countries and products the scenario which emerges from the European dried foods market is one of modest growth, a strong reliance on non-retail sales, most preferred by lower income families and generally sold in unexciting packaging.

GROWTH AREAS

Within the overall dried foods market there are several products and product categories which are experiencing rapid growth rates. In Europe the categories with the highest growth rates include

TABLE 7

Dried Food Market Growth

Product Category	Growth
Meat Substitutes	9.8
Complete Meals	4.9
New Breakfast Cereals	3.7
Fruit Juices	3.7
Potato Products	2.7
Soups	2.4

In the "instant" food and drinks category there are some particularly notable successes. Instant desserts and toppings are growing rapidly in the UK and have virtually doubled the market in recent years. Instant gravy granules have revitalized that market with annual demand growing by 25% in the UK. Instant soups are making rapid inroads internationally and are providing the main source of growth in the soup market. In the UK sales were up 30% in 1984 over 1983, mostly at the expense of non-instant dried soups. Instant drinks including chocolate and cocoa have also been successful and have given a new impetus to that sector. Low calorie products such as Carnation's sugar free instant hot chocolate have been highly successful, and low calorie cocoa now controls 30% of that segment of the US market.

Instant coffee dominates the coffee market although the traditional instant product is now being challenged by roasted and decaffeinated coffees. Instant potato has done well in the past but that market is tending to stabilize with the tremendous competition now offered by frozen products. Pot snacks and some other complete meals have had a checkered career, initially doing well when they entered the market but then fading away after a short period.

Health-conscious and dietary products have also shown significant growth in recent times. Apart from the low calorie drinks referred to earlier, there have been other remarkable successes in breakfast cereals and low calorie soups. Bran consumption, as a breakfast cereal, has grown rapidly and now holds over 12% of the UK market.

The success of these products shows that with the right product tailored to the consumer needs for quality and convenience, then dried foods can be highly successful. Given the emphasis on freshness and health by the modern consumer, Hilliam (ref. 5,6) concludes that "it might prove beneficial to manufacturers to market their products on a health and nutrition platform".

TECHNOLOGY AND TRENDS

TABLE 8

Most commonly used Food Drying Methods

Dryer Type	Typical Products
Spray Dryers	Milk/Coffee
Tunnel/Belt Dryers	Fruit/Vegetables
Through Circulation	Grain
Fluidized Bed	Sugar
Drum/Indirect Rotary	Pastes, Doughs
Freeze Dryers	Coffees, Mushrooms
Drum	Juices
Microwave	Chips

The trend appears to be towards compact plant where granulation, drying, heating and cooling treatment are all combined in one item of plant. Energy efficiencies are critical and heat pump technology should yield significant improvements in that field in future.

Karel (ref. 7) claims that the focus on quality retention during dehydration is only in its infancy but that its potential is enormous. It is most likely to be realised with the application of computers in food drying technology. If this is correct, and if the industry can improve the quality of its product to a significant degree and in a competitive manner, then the future prospect for the industry could be transformed.

QUALITY AND CONSUMER PERCEPTIONS

A commonly found theme in reports on the dried food market is the consumer's perception that dried foods are inferior to other forms of preserved foods. Karel (ref. 5) refers to the scientific explanations underlying this perception in highlighting the changes which occur to food during dehydration - "some of the most important are shrinkage, plasticization..... changes in diffusion co-efficients, and phase transitions which results in very extensive changes in the foods".

He is optimistic that researchers can improve the quality of dried foods with changes in technology and improved information on the topic.

Consumer research, however, highlights the enormity of the challenge for dried food manufacturers and marketers in convincing the consumer of the quality of dried foods. In 1985 housewives were asked to rate foods for freshness and the results are given in Table 9 (ref.2).

TABLE 9

Food Freshness on Scale 1 - 5

Food Category	Rating of 4 or 5 (%)
Frozen	72
Chilled	70
Unrefrigerated	64
Vacuum-Packed	64
Canned	40
Dried	32

Dried foods were ranked lowest, and apart from canned food were substantially lower than all other preserved food categories.

In a remarkably similar conclusion Mintel (ref. 2) commenting on the ready meal market note that "Quality equals taste in the consumer's view of ready meals. Chilled ready meals rate highly in this respect, closely followed by frozen ready meals. Canned ready meals are some way behind and dried ready meals are lowest".

Dried foods are also perceived as less convenient than frozen or chilled as they generally require more preparation for cooking. They also appear to be regarded as relatively expensive per serving and less nutritious.

Earlier in this paper reference was made to the fact that dried foods are preferred by lower-income families, have large non-retail sales and are often packaged in cheap and unexciting packs. These facts also adversely influence quality perceptions.

SOME IMPLICATIONS FOR IRELAND

Irish industrialists and researchers have a keen interest in international developments in such sectors as dried dairy products, soups, sauces, mixes, and vegetables.

In the dried milk sector, consumption is declining rapidly in the US and in Europe it is virtually static. The volume of exports from Europe in 1983 increased by over 8% while the value of these exports declined by 13% from 1982; a 21% decline in unit value over one year. The dependence on third world markets is increasing with 98% of the developed world exports of dried milk going there in 1983.

The Irish dairy sector is heavily dependent on the dried milk markets. The returns now available from intervention sales of dried products are generally insufficient to justify replacement of existing plant, and companies would need to exploit specialized niches in the market to earn adequate returns.

The Industrial Development Authority (IDA) would support market research and product development work to identify and provide suitable products for growing and high return niches in the dried milk market. It would also support fixed asset investment for the production of such products. Grants would not be provided, however, for replacing plant to produce purely commodity powders.

The markets for soups, sauces, and mixes are relatively buoyant, particularly for such products as instant soups and desserts, gravy granules and toppings. The IDA would evaluate individual projects in these areas on their merits and would be prepared to support suitable new projects or expansions to produce these products.

Within the dried vegetable sector, the US consumption of dried vegetables appears to be declining, and in Europe consumption is at best static. Existing sales are heavily concentrated in the industrial and institutional trades as an ingredient rather than a finished product. The freeze drying of vegetables has had a controversial background in Ireland and the difficulties of achieving a commercial success here for a branded product appears to be matched by similar difficulties elsewhere. This however, should not dampen enthusiasm for further Irish developments in this sector.

CONCLUSIONS

1. Demand prospects for dried foods in the developed world are poor for the immediate future. The market needs new products, a better image and heavy promotion to prevent a long term decline. Demand and needs are growing in the developing world but there are concerns about the commercial

viability of servicing these markets.

2. Within this overall gloomy trend, there are some products experiencing rapid growth. Good examples are "instant" food and drinks, together with health and dietary products. These growth opportunities are relatively small given the overall size of the dehydrated food market.

3. Dried foods are perceived by consumers as of inferior quality to other preserved foods. They are also seen as relatively expensive.

4. New drying technology has given a significant boost to the demand for dried foods in the past and some researchers feel that this can be repeated if more attention is given to food quality during drying. This is a major challenge for the industry.

ACKNOWLEDGEMENT

I would like to thank Dermot G. Gearty for his assistance in seeking out background data for this paper.

BIBLIOGRAPHY

1. Frost & Sullivan Inc., Dehydrated and Powdered Foods in Europe, Ref. E580, 1982.
2. Mintel Publications Ltd., Convenience Meals, May 1985, London.
3. Euromonitor Publications Ltd., The Convenience Foods Report, 1981, London.
4. Predicasts Inc., USA., Predicast Basebook, 1984.
5. Hilliam, Moira A, Instant Foods in the UK, Food Market Updates No. 7, Laboratories of the British Food Manufacturing Industries Research Association, November 1984, Leatherhead, Surrey.
6. Hilliam, Moira A, Low Calorie Foods in the UK, Food Market Updates No. 10, Laboratories of the British Food Manufacturing Industries Research Association, May 1985, Leatherhead, Surrey.
7. Karel, M, Advances in Food Dehydration, Internationale Zeitschrift fur Lebensmittel-Technologie und-Verfahrenstechnik 35(1) 6, (1984).

WATER ACTIVITY

CORNELIS VAN DEN BERG
Department of Food Science, Agricultural University Wageningen, The Netherlands.

ABSTRACT

During the last few decades, water activity has become a very useful tool in dealing with water relations of foods during processing. During that time, water activity has been studied in addition to water content, leading to many important developments in the dehydration, packaging and storage of foods.

In this paper, the meaning of water activity is derived from its thermodynamic background, and the importance of system equilibrium is emphasised.

Because micro-organisms respond differently to identical water activity levels set by different solutes, and because many foods are not in a state of equilibrium, as evidenced by hysteresis effects during humidification and drying, the use of water ac tivity concepts can not guarantee the accurate prediction of food shelf life.

The shortcomings of water activity in food processing can be overcome to some extent by the introduction of concept of the glass-rubber transition of foods.

Sorption isotherms of foods, i.e. the relationship between water content and relative vapour pressure, and the factors affecting them, are also briefly considered. Developments in measurement of vapour pressure and sorption isotherms are discussed, and methods of calculating water activity from food composition are also surveyed, both for solutions and for foods with undissolved solids.

INTRODUCTION

Water is the single most important factor governing food deterioration processes, of which microbial spoilage is usually the most rapid. The oldest methods used for preservation of food, namely concentration, drying, smoking, and salt and sugar infusion, depend on the principle of reduced water content. However, it is not just the water content which determines the shelf life of food, but rather some other factor related to the availability of water for decay reactions. For instance, peanut oil spoils slowly at 0.6% moisture, crystalline white sugar with 2% moisture is unstable, while potato starch is stable at 20% moisture.

The question of water availability is a complex one, and it depends on the many different forms that constituent water in foods can take. An appreciation of the factors affecting the state of water in foods is important to the understanding of water availability. There is no single parameter which can be used as a reliable guide to predicting food spoilage or to determining the drying endpoint required for a shelf stable product. The most useful such parameter is water activity (a_w), and much progress has been made in the water relations of foods since water activity research began. Many new drying and storage procedures have been devised, and new products such as intermediate-moisture foods ($0.65 < a_w < 0.95$) have been developed.

The relationship between water content and water activity or relative vapour pressure is known as the water sorption isotherm, and this concept is used to establish product specifications for the effective drying, packaging and storage of foods.

Literature on water activity has proliferated over the last two decades. While this paper is not a comprehensive survey of water activity, it focuses on the strengths and weaknesses of the concept, the methods available for its measurement, and developments in its application to foods. Theoretical aspects of water activity are emphasised since in the opinion of the author some water activity data which is being determined experimentally could be more easily determined by calculations with at least equal accuracy.

BACKGROUND

As far as the author is aware, water activity of foods was first mentioned in the literature in 1949 when Mossel and Westerdijk (ref. 2) correctly described water activity as a characteristic of the substrate, and as one of the essential factors governing the spoilage of food by micro-organisms, even though they had not developed a theoretical basis for water activity at that time (Mossel, personal communication). In 1953 Scott (ref. 3) related relative vapour pressure of foods to the thermodynamic activity of water, and showed a clear correlation between water activity and growth of Staphylococcus Aureus. A review of the topic by Scott in 1957 (ref. 4) found general acceptance and stimulated a rapid expansion in theoretical developments and application of a_w.

Many studies were reported on the relationship between relative humidity and the deterioration of biological products, before water activity began to be studied. Measurements of sorption isotherms for textile fibres were reported as long ago as 1893 (ref. 5). Chemical engineers involved in drying operations have long been aware of the relationship between water content and vapour pressure, and of the role of vapour pressure differences as a driving force in dehydration. In 1924, Walter (ref. 6) reported that for effective preservation of biological materials by dehydration, their water content should be reduced so that relative vapour pressure is less than 85%.

DEFINITION OF WATER ACTIVITY

Water activity is a thermodynamic concept, defined as the ratio of fugacities, and a brief review of basic and applied thermodynamics is essential to its understanding.

Any system in equilibrium is characterised by a minimum of free energy. When a system such as food undergoes a change, it is usually at constant total presure. Its equilibrium is governed by the Gibbs free energy (G) also known as free enthalpy or Gibbs function. For a closed system with no exchange of mass between surroundings and environment, the function is defined as

$$G = U + pV - TS = H - TS \tag{1}$$

where U = total internal energy, p = pressure, V = volume, T = absolute temperature, and S = entropy (a measure of disorder in the system. (U + pV) is known as enthalpy (H). In liquid and solid phases, pV is negligible at ambient temperatures.

The Gibbs function can be applied to a pure substance, or to one component of a mixture. For any component i in a mixture, the Gibbs free energy expressed on a molar basis, is given by

$$\mu_i = \left[\frac{G}{n_i}\right]_{p,T,n_i} \qquad \text{(mixture)} \tag{2}$$

where n = the number of moles, μ_i = the partial molal Gibbs free energy for component i, and is the differential change in Gibbs free energy of the system with the addition of one mole of pure component i to the system at constant temperature, constant pressure, and constant quantities of other components (n_j) j ≠ 1. (ref. 7). When a system consisting of various phases and components is in thermodynamic equilibrium, molecules are interchanging continuously between phases. The rates of interchange of molecules of all species between the phases are equal and there is no net change in the mass of the various phases. Hence, there is no net change in free energy between the phases, and the chemical potentials of all components within the phases are equal.

The chemical potential (μ) is related to another property named the absoute activity, (μ), by the following relationship (ref. 8).

$$\mu = RT \ln \lambda \qquad (3)$$

where R denotes the ideal gas constant.

The absolute value of the chemical potential is not known, and the absolute activity of any component in a mixture must be related to the activity of that component in a reference or standard state, which for convenience, is taken as pure component at the same temperature and total pressure. Hence, equation (3) now becomes:

$$\mu_i - \mu_i^o = RT \ln \frac{\lambda_i}{\lambda_i^o} = RT \ln a_i \qquad (4)$$

The superscript^o denotes the standard state, i.e. pure component i. The ratio of absolute activities (a_i) is called the relative activity of component i. When component i is water, the ratio of absolute activities is known as water activity. Due to dilution and molecular interactions, the activity of a pure substance is always lowered upon mixing with other molecular species, and the value of the relative activity of any substance in a mixture always lies between 0 and 1.

In applying the theoretical concept of activity to real systems the chemical potential and absolute activity cannot be determined directly. Hence they must be expressed in terms of an auxiliary function, namely <u>fugacity</u>, which is the key to the practical use of water activity. For <u>ideal</u> systems, the chemical potential can be related to vapour pressure. By combining the ideal gas law with the well-known Gibbs-Duhem equation at constant temperature, the following relation is obtained:

$$d\mu_i = RT \, d \ln p_i \tag{5}$$

where p_i denotes partial vapour pressure of component i. To generalize this relation for real systems, Lewis in 1901 defined fugacity as a corrected pressure function (ref. 9). For a real gas:

$$f = p \exp \frac{\mu(T,p) - \mu^{id}(T,p)}{RT} = p \frac{\lambda}{\lambda^{id}} \tag{6}$$

where the superscript id denotes an ideal gas (ref. 10). Fugacity has the dimensions of vapour pressure, and represents the tendency of the molecules of a substance to escape. For an ideal gas in a mixture, it is identical to partial pressure. For real gases, as total pressure decreases, fugacity approaches vapour pressure, since all gases exhibit ideal behaviour at zero pressure.

$$\frac{f_i}{p_i} \rightarrow 1 \text{ when } p \rightarrow 0 \tag{7}$$

Water has a low vapour pressure at ambient temperatures and so non-ideal behaviour in the vapour phase can be disregarded and the activity of water in a real system becomes:

$$a_w = \frac{f_w}{f_w^o} \approx \frac{p_w}{p_w^o} \tag{8}$$

Hence, the water activity of a system is closely equivalent to the relative vapour pressure (RVP) of water over that system, although it is not identical with it in thoery. Gal (ref. 11), who investigated the numerical difference between a_w and RVP, showed that at ambient conditions (below 50°C, at pressure of 1 bar) the difference between the two is less than 0.2%. This difference may be neglected for practical purposes, and justifies what is often stated; viz: that water activity equals relative vapour pressure.

Since pure water is taken as reference state, it follows that for subfreezing temperatures, all systems containing ice have the water activity of ice at the temperature of the system. The water activity of ice is defined by:

$$a_w\ (\text{ice}) \approx \frac{p_{ice}}{p_w^{\,o}\ (\text{supercooled})} \tag{9}$$

where $p_w^{\,o}$ is the vapour pressure of supercooled water at the temperature of the ice. Vapour pressures for supercooled water are known down to about -36°C. Hence, ice is formed in a system only when the system a_w is higher than the ice a_w at the freezing temperature. An application of this principle is the recent development of "freeze-flow" intermediate moisture food products that can be used directly from the freezer because they are largely free of ice at low temperatures.

The vapour pressure of water in equilibrium with food can be measured. Equilibrium relative humidity (ERH) is 100 times RVP:

$$a_w \approx \frac{p_w}{p_w^{\,o}} = \frac{ERH}{100} \tag{10}$$

It must be remembered that since a_w is defined in this way, it can strictly be applied only to a system in thermodynamic equilibrium, and its value is true only for specified conditions of temperature and total pressure. However, for most food systems, true thermodynamic stability, constant temperature, and constant pressure are seldom realised in practice.

FOOD STABILITY

Next to temperature water activity is now considered as probably the most important parameter in food preservation and processing. Table 1 shows the wide variation in water activity values for foods.

TABLE 1
Average values of water activity a_w (expressed as relative vapour pressure) and percentage water (W) (wet basis) for some foods:

	a_w	W (%)
Milk	0.994	87
Fresh meat	0.985	70
Bread	0.96	40
Old cheese (Gouda)	0.94	40
Sausage	0.89	42
Marmalade	0.86	35
Sugared condensed milk	0.83	28
Wheat flour	0.72	14.5
Raisins	0.60	27
Marcaroni	0.45	10
Boiled sweet	0.30	3
Biscuits	0.20	5
Skim milk powder	0.11	3.5
Crisp potato chips	0.08	1.5

While water activity is a useful concept in food preservation, there are limitations to its application in practice, especially where foods exhibit non-equilibrium behaviour. Most foods and related products are multi-component systems consisting of many phases. At higher moisture levels, the components are partially dissolved and may exist as gels. The undissolved components usually constitute the bulk of the total dry solids in the food system. These are relatively high-molecular-weight materials, polysaccharides and proteins, which form the structural framework in which the liquid phase is contained, either dispersed or continuous. When oils or fats are present, even more phases exist. The structure of these systems is difficult to characterize. When dried, the situation is further complicated by the effects of delayed crystallization or delayed dissolution. At low moisture levels the mobility of components is greatly diminished, and the mainly amorphous macromolecules are probably in glass condition. In the long term all amorphous substances tend to shift towards a more crystalline thermodynamically stable state, e.g. amorphous sucrose (refs. 13, 14) and amorphous lactose in spray-dried milk powder (ref. 5). Because many crystals bind either no water or less than in the amorphous state, the product vapour pressure shifts towards higher values on crystallisation. In food systems, hysteresis between drying and humidification is often observed. In a recent study on water-sorbent-solute relationships, Johnston (ref. 16, 17) observed important discrepances in vapour pressure behaviour for identical mixtures prepared in different ways.

It can be seen from the above that many foods systems are not truly stable but are rather in a state of pseudo-equilibrium or steady state. Hence in these instances the term "water activity" does not apply, but rather the term relative vapour pressure or relative humidity. Only when the system exhibits no hysteresis, can one be sure that vapour pressure is a measure of activity.

Nevertheless, in food science over the last two decades the concept water activity has been widely used and often misused. For products with low equilibration rates, the term "superficial water activity" is sometimes applied (ref. 18).

Another area where the limitations of water activity are not always correctly understood is in the interpretation of the food stability water activity relationship (fig. 1).

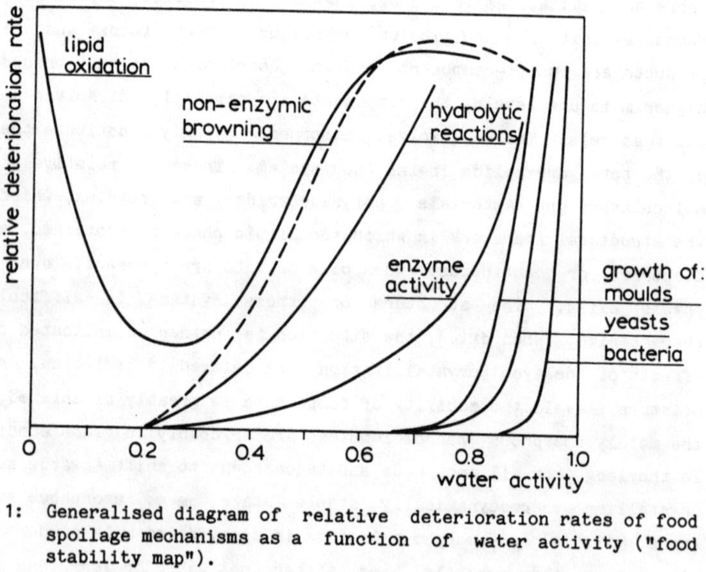

Fig. 1: Generalised diagram of relative deterioration rates of food spoilage mechanisms as a function of water activity ("food stability map").

This diagram by Labuza (ref. 19), relates many common food deterioration reactions to water activity, using results of individual observations of decay reactions in different food systems. It is a composite of trends that often vary from product to product, due to the variability of foods.

Nevertheless, it has been suggested that this diagram can be used to define safety limits for the spoilage of foods in general. This is a misapplication of water activity concepts, since the effect of water activity in foods depends on the composition of the solute. For instance, Corry (ref. 20) reported that growth of a Pseudomonas species stopped at $a_w = 0.97$ when adjusted with sodium chloride, at 0.964 with sucrose and at 0.945 with glycerol. Beuchat (ref. 21) also reported different microbial reactions at comparable a_w - values adjusted with different solutes. Apparently the microbial cell is not just a simple osmometer that stops working as soon as a certain osmotic pressure has been attained.

It is not surprising therefore that in recent years misconceptions have led to some difficulties in the preservation of intermediate moisture products. New products were formulated based on acquired experience, setting safety margins on the basis of a_w. The new product, however, although having the same a_w as its safe predecessor but with a slightly different humectant composition, spoiled on the shelf by microbial action. In fact, water activity is just one of many factors affecting microbial growth, and plays no more important a role than temperature, pH, or substrate composition. Hence, although Scott in his acclaimed papers (refs. 3, 4) was aware of the theoretical background of water activity, he did not distinguish clearly enough between product RVP and thermodynamic a_w.

It can be concluded in regard to figure 1 that it is more appropriate to make a clear distinction between the equilibrium nature of water activity, and the kinetics of deterioration reactions. In these reactions, water often plays a complex role. It may act as a reactant, a solvent or a secondary influence on the nature of the system. In practice, conclusions with regard to safe and economic specifications for dehydration and storage of a specific product should be drawn up only after careful consideration of the relevant water relations and conducting shelf-life studies.

FOOD-WATER INTERACTIONS

There is a strong interaction between water and dry foods and in particular between water and the polymer components that are usually highly polar in nature. In general the higher the molecular weight the greater the strength of the interaction between an individual polar residue and an associated water molecule (ref. 22, 23). The influence of these macromolecules on a_w begins at water contents less than about 30 to 45% (wet basis) depending on the macromolecule involved, and increases progressively as the water content decreases.

The following types of interaction occur at sub-microscopic level between water and different food components: (i) London-van der Waals dispersion forces, (ii) specific hydrogen bonds, (iii) Coulomb forces between water, ions and dissociated groups, (iv) steric effects, (v) dissolution effects (water soluble components), (vi) changes in mobility of polymer segments and other amorphous compounds caused by the glass-rubber transition, and (vii) capillary forces.

The first five mechanisms are commonplace and need no further comment. The glass-rubber transition (vi) is less well known. For pure amorphous compounds this secondary transition occurs upon heating at roughly 0.7 of the melting temperature (Kelvin). For polar compounds the temperature of this transition is influenced by the action of water as a plasticizer. During the transition the initially glassy state is changed to a tough rubber condition due to increased molecular mobility. At low water contents, the mobility of many food components is influenced by increasing moisture in much the same way as by increasing temperature. Figure 2 illustrates the glass-rubber transition for a starch-water system.

Food technologists could apply this principle more systematically than at present, because in the glassy state food products are stable for a long period of time, and because the diffusivity of components is several orders of magnitude lower in the rubber states. Below the glass-transition temperature (T_g), which for a mixture may well be a range of temperatures, water becomes unavailable for deterioration reactions within usual food storage periods. The effect of RH on the crystallization rate of amorphous sucrose as reported by Karel (ref. 13) is an example of this. Another example was recently reported by Slade (ref. 1) who observed that the defect ice grain growth did not occur in samples of ice cream stored below T_g. Collapse temperatures during freeze-drying can also be explained in this way (ref. 24).

It is often assumed that capillary forces (vii) depress vapour pressure in food systems that contain capillaries. This mechanism, called capillary condensation, is quantified by the Kelvin equation. However, it can be shown (ref. 25) that in gel-like systems such as starches, no significant capillary condensation occurs. Due to the effect of swelling pressure of water in small capillaries, this mechanism acts only in rigid capillaries, and not in swelling food poly-gels where capillaries act as additional barriers to water diffusion during drying.

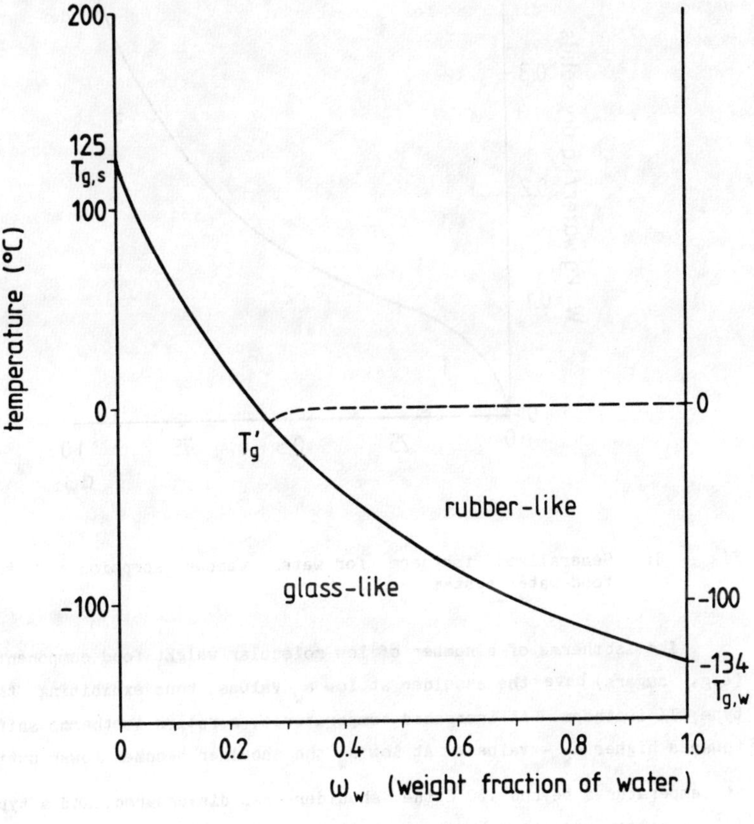

Fig. 2: The approximate glass-rubber transition temperatures as a function of mass fraction for the starch-water system.

SORPTION ISOTHERMS

Sorption isotherms of food products are generally sigmoidally shaped (fig. 3), being type II of the BDDT isotherm classification (ref. 26).

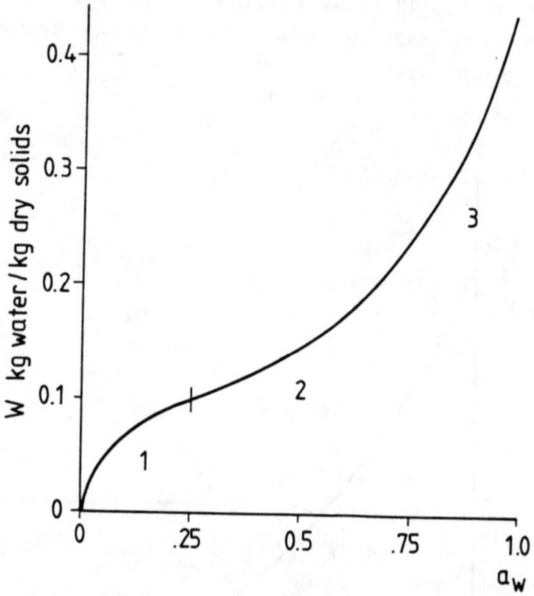

Fig. 3: Generalized isotherm for water vapour sorption of the food-water system.

The isotherms of a number of low molecular weight food components (e.g. sugars) have the shoulder at low a_w values, thus exhibiting the type III isotherm. At increased temperatures, sorption isotherms shift towards higher a_w - values. At low a_w the shoulder becomes lower until at temperatures beyond 100°C the shoulder has disappeared, and a type III isotherm results. Numerous studies carried out over the last few decades support the view that most water molecules sorbed at low a_w values (range 1 in figure 3) are strongly held with greatly restricted mobility, behaving almost like part of the solid. This water is sorbed at polar groups of the food solids that act as rigid physical adsorbents. It is often referred to as "monolayer" water, arising out of isotherm analysis using the BET (ref. 27) or comparable isotherm equations. For starches for example it is found that the monolayer value thus derived is approximately one water molecule per anhydroglucose monomer (refs. 25, 28).

Water in the steep part of the isotherm (range 3 in figure 3) is more or less free water, mechanically entrapped in the voids of the system, with properties similar to those of bulk water. Here the system behaves as a typical gel. A high water uptake at these a_w values indicates that the major components of the system are dissolving. In range 2 (fig. 3) water is in transition between states 1 and 3. Here water-dependent chemical and biochemical reactions start to take place. The amorphous structure is plasticized and its volume increases in proportion to the uptake of water.

One difficulty with using sorption isotherms for solid food systems is that due to the nature of food-water systems, the exact position and shape of the isotherms depend on a large number of factors. Of these, chemical composition, temperature, physical structure (e.g. amorphous or crystalline) are obvious, but origin (harvest) and pretreatment of the sample (e.g. the way it was dried), methodology, desorption or resorption (hysteresis) can also play a role, and are less predictable. This can make it difficult to reproduce the isotherm exactly. The difference between the first desorption isotherm of a newly harvested product and the one measured after some time, when the product has "set", is well known. For this reason, when accurate isotherms are required, sorption isotherms by other workers should not be used without good reason, and only if all relevant data are reported. Compilations of food isotherms (ref. 29, 30, 31) are also of limited value, except as a first estimate. The wide scatter in literature values of isotherms for various food products is reported in the comprehensive treatise of Heiss (ref. 29)

That accurate determination of sorption isotherms requires a high level of skill was convincingly illustrated in a collaborative study organised by Multon (pers. comm.). Resorption isotherms of identical samples of native cassava starch were measured at 25°C in eleven laboratories. The a_w results varied by $\geq$ 0.2 and 4% in water content (dry basis) in the middle of the curve. A collaborative study organised by the EEC-COST subgroup on a_w (refs. 32, 33) using common test methods, achieved better results.

Hysteresis is commonly observed between equilibria established by desorption (drying) and resorption (humidification) and often extends over the full isotherm. Typical food gels, e.g. starch or protein gels, tend to close the hysteresis loop beyond $a_w \approx 0.9$, indicating that these systems are solutions. The phenomenon is related mainly to the polymer components of the food system, it is influenced by most changes in system conditions, and is difficult to reproduce in successive cycles of desorption and resorption. The widest hysteresis loop for a sample at one temperature is observed when measurements are begun from both ends of the humidity range, but almost any point within the loop can be attained by scanning. The influence of temperature on hysteresis is large. At low temperature the hysteresis effect is increased, and is diminished as temperature increases, until at about 80°C no hysteresis effect is found. This is illustrated in fig. 4 which shows resorption and desorption isotherms at 5, 25 and 45°C for milled wheat. At 45°C the hysteresis effect is almost negligible.

MEASUREMENT OF WATER ACTIVITY

Two approaches are used for the measurement of water activity or more generally the sorption isotherm (a_w and water content), viz. gravimetric and manometric or hygrometric methods. Gravimetric methods involve taking sample measurements after equilibration, while in the manometric or hygrometric technique measurements are taken from the atmosphere in equilibrium with the sample. The principles and instrumentation of these methods have been described comprehensively by Gal (ref. 34).

Gravimetric methods are generally more accurate but more time consuming compared to hygrometric methods and are therefore often preferred for research. A variety of sensitive balances with static or dynamic equilibration procedures has been built over the years (refs. 35, 36, 37), but the simpler low-cost but laborious method using dessicators with saturated salt solutions to maintain constant humidity is still most widely used. This was the method used in the EEC-COST group collaborative study (ref. 32) and researchers beginning work in this area should adopt this method and begin with the reproduction of the standard isotherm for micro-crystalline cellulose produced by the COST group.

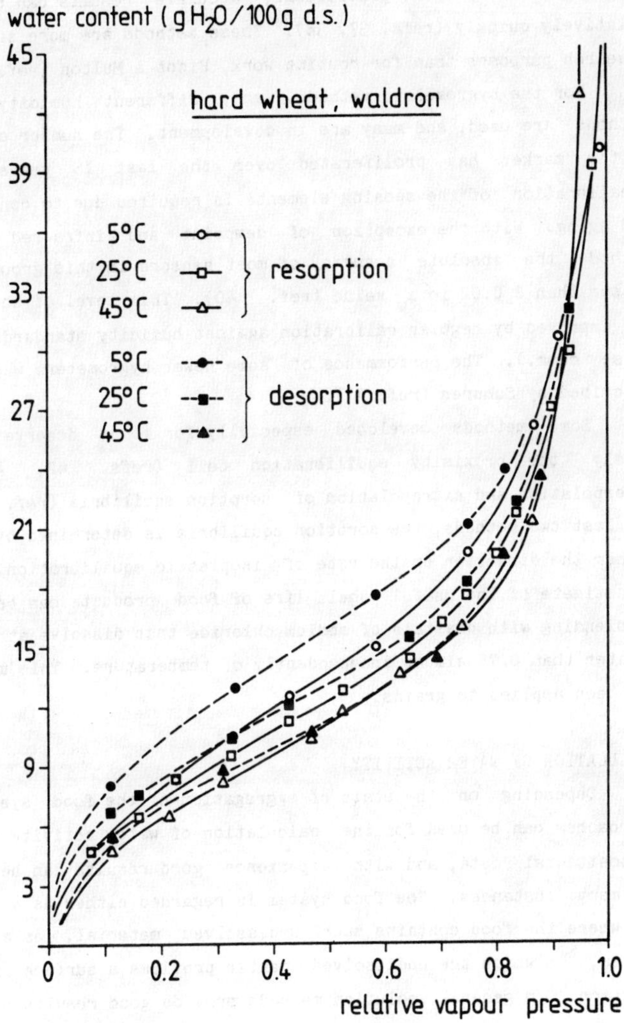

Fig. 4: Hysteresis effect between desorption and resorption at three temperatures, as observed for milled hard wheat, type waldron origin USA.

In manometric methods the vapour pressure of product is measured directly. With proper precautions, accurate results can be obtained relatively quickly (refs. 37, 38). These methods are more suitable for research purposes than for routine work, Bizot & Multon (ref. 39).

For the hygrometric methods, many different humidity measuring methods are used, and many are in development. The number of devices on the market has proliferated over the last 15 years. Regular recalibration of the sensing elements is required due to contamination and aging. With the exception of dewpoint and infra-red absorption methods the absolute accuracy of most sensors in this group is not better than ± 0.02 in a_w value (ref. 40). This level of accuracy can be improved by regular calibration against humidity standards (Ruegg, pers. comm.). The performance of some newer hygrometers was recently described by Schurer (ref. 41).

Some methods developed especially for foods deserve mention, namely the proximity equilibration cell (refs. 42, 43), and interpolation and extrapolation of sorption equilibria (ref. 44). In the last two methods, the sorption equilibria is determined by scanning either the direction or the rate of isopiestic equilibration. Finally an estimate of the useful shelf life of food products can be obtained by blending with crystals of sodium chloride that dissolve at a_w values greater than 0.75 almost independently of temperature. This method has has been applied to grains.

CALCULATION OF WATER ACTIVITY

Depending on the state of aggregation of the food system, two approaches can be used for the calculation of water activity from food compositional data, and with experience good results can be obtained in many instances. The food system is regarded either as a solution or where the food contains much undissolved material, as a sorption system, in which the undissolved matter provides a surface for water sorption. In general, solution methods provide good results at high a_w, and sorption system methods good results at lower a_w.

Ideal solutions

With ideal solutions, pure components can be introduced into the system without change in volume, heat or entropy. Raoult's law applies:

$$a_w = x_w = \frac{n_w}{n_w + n_s} \tag{11}$$

where x_w is mole fraction of water; n_s and n_w are the number of moles of solute and water respectively. Equation (11) accurately predicts a_w for dilute aqueous solutions of low-molecular solutes such as sugars (sucrose < 2 molar; glucose and fructose < 4 molar) and salts (< 1 molar, in which dissociated ions are counted as molecules).

Non-ideal solutions

Departures from ideality proceed from interactions both between different and identical molecules in the mixture. These interactions can arise from molecular forces as well as from differences in molecular size. Raoult's law in its non-ideal version reads:

$$a_w = \gamma_w x_w \tag{12}$$

where γ_w is the activity coefficient of water in the system, and represents the non-ideal portion of the absolute activity for the real component.

In the field of solution theory, assumptions regarding molecular interactions have led to a number of theories and equations which have not yet been fully explored for possible use in predicting a_w (ref. 12). One such equation, the one-constant Margules equation, was used successfully by Taylor and Rowlinson (ref. 45) to determine the activity coefficients for water in glucose and sucrose solutions.

This equation is now known as the Norrish equation (13) (ref. 46).

$$\log_{10} \gamma_w = \log_{10} \left(\frac{a_w}{x_w}\right) = k\, x_s^2 \tag{13}$$

gives accurate values of a_w for single solutions of many sugars, organic acids and other organic molecules (refs. 46, 47, 48). k is a constant for binary mixture in question. The equation was generalized as follows for multi-component solutions, and found to fit simple sugar mixtures (ref. 46).

$$\log_{10} a_w = \log x_w - (k_2^{\frac{1}{2}} x_2 + k_3^{\frac{1}{2}} + k_4^{\frac{1}{2}} x_4 + \ldots\ldots)^2 \qquad (14)$$

where k_2, k_3, etc are the constants for the binary mixtures. However, for more complex mixtures it is easier to average the individual component contributions, returning essentially to equation (13) (refs. 46, 48). When equation (13) came into use it superseded earlier empirical equations for sugar solutions (refs. 49, 50).

Values for a_w of high moisture mixtures (a_w > 0,75) can be predicted using the Ross equation (ref. 51):

$$a_w = a_{w,1}\, a_{w,2}\, a_{w,3} \ldots\ldots \qquad (15)$$

where $a_{w,1}$ is the a_w of a solution of component 1 dissolved in the total amount of water present. Ross derived this equation from the Gibbs-Duhem equation, using certain simplifications. He claimed the equation to be valid for solids also, but in fact when solids are present in substantial amounts, errors are caused (refs. 47, 48, 52, 53). Equation (15) is essentially a solution equation.

A modification of the Ross equation was proposed by Ferro Fontan and Chirife (ref. 54) as follows:

$$a_w = [a_{w,1}\,(m)]^{\frac{m_1}{m}}\, [a_{w,2}\,(m)]^{\frac{m_2}{m}}\, [a_{w,3}\,(m)]^{\frac{m_3}{m}} \ldots\ldots \qquad (16)$$

where $[a_{w,1}\,(m)]$ is the a_w of a solution of component 1 at the total molality m of the mixture, and m_1/m is the ratio of the molality of component 1 to the total molality. This equation predicts a_w more accurately than equation (15) but its use requires more information on the components.

Because of their electrical charge, electrolyte solutions need a different approach (ref. 56). Equation (16) is applicable to dilute electrolyte solutions if molality is replaced by ionic strength (ref. 53), while other methods (refs. 57, 58, 59) yield better results over a wider range of concentrations. Salts, however, when not present in considerable quantities for preservation purposes, have a relatively small influence on a_w - usually less than a few percent.

Differences in the size of solute and solvent molecules cause non-ideality in the entropic properties of the solution. Lilley (ref. 1) showed significant departures from ideality when the ratio of molar volumes of solute and solvent is of the order of 500 or more. This is the case for polymer solutions, for which Flory and Huggins derived their solvent activity equation (ref. 60):

$$a_w = (1 - \phi) \exp [(1 - \frac{V_w}{V_p}) \phi + \chi\phi^2] \qquad (17)$$

where ϕ is the volume fraction of polymer, χ is the Flory-Huggins interaction parameter and v_w and v_p are partial molal volumes of water and polymer respectively. This equation was found to describe satisfactorily the sorption isotherm of gelatinized starches for a_w > 0.9 (refs. 28, 61).

Activity is a colligative property: i.e. it depends primarily on the number density of solute particles, and therefore has a direct relationship to other colligative properties such as osmotic pressure, swelling pressure, boiling point elevation and freezing point depression. At the freezing point the following equation applies (ref. 48):

$$-\ln a_w = 0.0096934\ \Delta T_f + 0.000004761\ \Delta T_f \qquad (18)$$

where ΔT_f is the freezing point depression (°C). Since freezing point determinations can be carried out very accurately, this method can be recommended, provided it is realized that the result is the a_w at the freezing point. For extrapolation to other temperatures, the influence of temperature on a_w for the solution must be known. Ferro Fontan et al. (ref. 48) successfully applied this method to sugar solutions, as they have small temperature coefficients of a_w.

Undissolved systems

To predict sorption equilibria for foods containing undissolved solids, the experimental curves must be described mathematically. For practical applications such as drying, packaging and storage, it is important to have a relatively simple equation for this purpose, with a limited number of parameters. These parameters should preferably have a physical significance and known temperature dependence, and if possible, should be able to be corrected for the influence of hysteresis. A 1981 survey (ref. 12) showed that about seventy-five different mathematical equations were proposed to describe food-water isotherms, and since that time others have been added (refs. 62, 63), which illustrates both the importance and the difficulty of the problem. The equations range from theoretical to empirical, and use from one to seven parameters. In addition to the one-parameter equation (17) of Flory and Huggins, three others will be discussed here.

Of theoretical interest is the sorption model of Tomka (refs. 64, 65), developed especially for swelling biopolymers with different polar groups, consisting of a set of three equations with up to seven parameters. These fit the sorption isotherm of casein with good accuracy. Although this sorption model is of high quality, it is not particularly useful in practice due to its complexity and large number of parameters.

For sorption isotherms below 0.4 a_w and for the determination of specific surfaces, the BET two-parameter equation (ref. 27) is still generally used. The monolayer water content calculated using the BET equation is sometimes considered as a condition of maximum stability for several food products, which is in agreement with the data of figure 1.

In recent years the more widely applicable GAB equation (19) is gaining ground for the analysis and description of food sorption isotherms (refs. 25, 66, 67, 68, 69, 70). It is derived independently by Guggenheim, Anderson and de Boer -GAB- (refs. 71, 72, 73) from a physical adsorption model related to the BET theory. The GAB theory has been described in detail elsewhere (ref. 28). It contains aspects of solution theory as well as physical adsorption, and may have wide application. The GAB equation reads:

$$W = \frac{W_m C_G K a_w}{(1 - K a_w)(1 - K a_w + C_G K a_w)} \tag{19}$$

where W is the mass fraction of water on a dry basis, W_m is the mass fraction of water at monolayer coverage, and C_G and K are interaction energy constants, related as follows:

$$C_G = c \exp\left(\frac{H_1 - H_2}{RT}\right) \qquad K = k \exp\left(\frac{H_l - H_2}{RT}\right) \tag{19 bis}$$

where c and k are entropy-related constants, H_1 and H_2 are the enthalpies of sorption for the first and further sorbed layers of molecules, and H_l is the enthalpy for bulk water (enthalpy of condensation).

The BET equation is a limiting case of the more general GAB equation for K = 1. Equation (19) can be used to describe with good accuracy the majority of food isotherms up to about 0.9 a_w. It also compares very favourably with other equations for the description of isotherms (ref. 68). Hence, it is concluded that the GAB equation is probably the simplest and most widely applicable isotherm equation available.

A combination of equations (17) and (19) with only four physically significant parameters was used to describe accurately the full isotherm of gelatinized starch (ref. 28). Hence, only four measurements, well distributed over the total range, are required to determine the full isotherm. However experimental evaluations should be done at regular intervals in view of the variability of natural products.

Procedures for dry mixtures

For a mixture of dry components, a_w is calculated from the sorption isotherms of the components in proportion to their mass ratios. Salwin and Slawson (ref. 74) used this principle to develop their approximation method, in which portions of the individual isotherms are linearized and a_w is calculated from the mass fraction, isotherm slope and a_w of the individual components. This approximation gives good results if major hysteresis effects are absent.

To determine the water activity of combinations of moist and dry components at high humidity, the Ross equation (15) can be used, but accuracy is variable.

Influence of temperature

Since food components do not usually behave as ideal mixtures, a_w changes with temperature. The effect of temperature is of great practical importance. For instance, grain with 15% water which is stable in Northern Europe, spoils in tropical Africa where the maximum allowable water content is 13.5%. However, relatively few studies have been done in this area. The effect of temperature on sorption has been thoroughly reviewed by Weisser (ref. 70) and does not need to be repeated here. Estimating the effect of temperature using the GAB equation (19) gives correct values over limited ranges of temperature. To increase the temperature range, Weisser related the monolayer value W_m to temperature using an Arrhenius-like relationship. However, while this approach seems useful in practice, its theoretical basis is doubtful.

The best mathematical description of the effect of temperature on a_w is that of Riedel (ref. 75). Based on his extensive and accurate measurements of the enthalpy of sorption of many products, a standard enthalpy curve as a function of water content was obtained. This curve was combined with the well-known Clausius-Clapeyron equation to yield:

$$\ln \left(\frac{a_w(T_2)}{a_w(T_1)}\right)_W = A\, e^{-BW} \left(\frac{1}{T_1} - \frac{1}{T_2}\right) \qquad (20)$$

where indices 1 and 2 signify the two temperatures under consideration and A and B are product-specific constants. This equation which has not been extensively tested relates a_w and T in almost the same way as the Clausius-Clapeyron equation. Due to the way the function A exp(-BW) is derived, equation (20) is expected to be at least as accurate as the Clausius-Clapeyron equation.

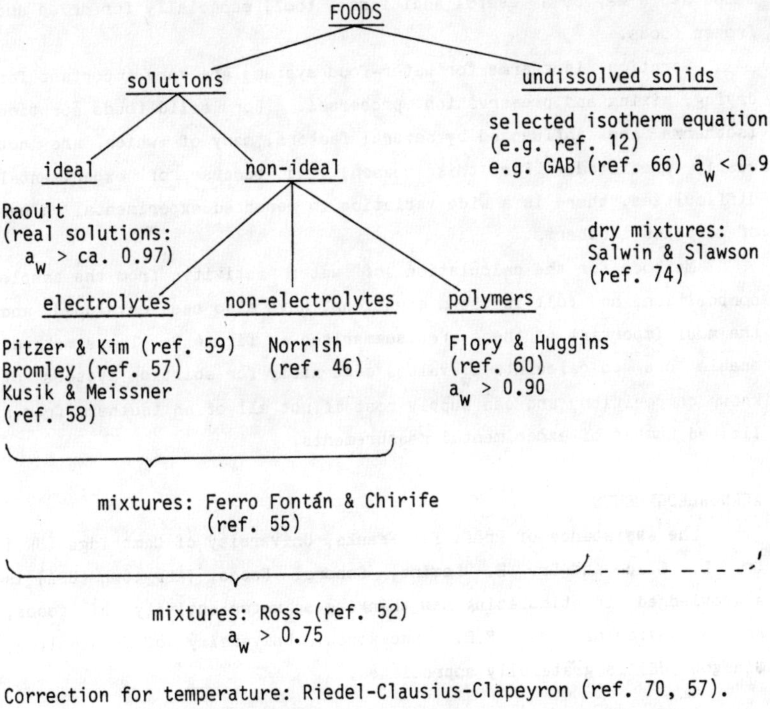

Fig. 5: Survey of methods for the calculation of water activity or relative vapour pressure of foods.

CONCLUSION

During the last three decades water activity defined in terms of vapour pressurehas become very important in food science and technology as an indicator for the availability of water for spoilage reactions. However, the interpretation of the concept in relation to food deterioration has not always been correct. From the definition of a_w it follows that the concept applies strictly only to systems in thermodynamic equilibrium. Foods are seldom in equilibrium but rather in a steady state, in which case the notion water activity should not be used, but rather relative vapour pressure or relative humidity.

Due to the variability of foods shelf-life predictions cannot be made solely on the basis a_w. The concept of glass transition temperature may be a useful additional tool, especially for dried and frozen foods.

Sorption isotherms for water-food systems are very important for drying, mixing and preservation processes. For solid foods sorption isotherms are influenced by several factors, many of which are not easily predicted. For this reason, and because of experimental difficulties, there is a wide variation in reported experimental values of sorption isotherms.

Methods for the calculation of water activity from the sample composition, and full isotherm equations, have also been developed, and the most important of these are summarised in figure 5. These methods enable one to calculate a_w values accurately for solution systems of known composition, and can supply most if not all of an isotherm from a limited number of experimental measurements.

ACKNOWLEDGEMENTS

The assistance of Prof. F. Franks, University of Cambridge (UK), Dr. L. Slade and Dr. H. Levine, General Foods, Tarrytown (USA) is acknowledged in stimulating new thinking on water activity of foods, and the advice of Dr. R.B. Duckworth, University of Strathclyde, Glasgow (UK) is gratefully appreciated.

REFERENCES

1 F. Franks, L. Slade and H. Levine (eds), Water activity: a credible measure of technological performance and physiological viability? Proc. discussion conf., Cambridge (UK), July 1-3, 1985 (to be published).
2 D.A.A. Mossel and J. Westerdijk, Anthonie van leewenhoek,, 15(1949)190.
3 W.J. Scott, Australian J. Biol. Sci., 6(1953)549.
4 W.J. Scott, Adv. Food Res., 7(1957)83.
5 M.T. Schloesing Jr, Bull. Soc. Encour. Ind. Nat., 92(1893)717.
6 H. Walter, Z. F. Botanik, 16(1924)353.
7 J.M. Prausnitz, Molecular thermodynamics of fluid phase equilibra, Prentice Hall, Englewood Cliffs N.J., 1969.
8 G.N. Lewis, M.E. Randall, K.S. Pitzer and L. Brewer, Thermodynamics, 2nd rev. ed., McGraw-Hill Inc, New York, 1961.
9 G.N. Lewis, Z. Phys. Chem., 38(1901)205; Proc. Am. Acad. Sci., 43(1907)259; Z. Phys. Chem., 61(1907)129.

10 S.I. Sandler, Chemical and Engineering thermodynamics, J. Wiley & Sons, New York, 1977.
11 S. Gal, Helv. Chim. Acta, 55(1972)1752.
12 C. van den Berg and S. Bruin, in L.B. Rockland and G.F. Stewart (eds) Water activity: influences on food quality, Academic Press, New York, 1981, p. 1.
13 M. Karel, in A. Spicer (ed), Advances in preconcentration and dehydration of foods, Applied Science Publ., London, 1974, p. 45.
14 D. Roth, Doctoral thesis, Universitat Karlsruhe, 1976.
15 E. Berlin, B.A. Anderson and M.J. Pallansch, J. Dairy Sci., 51(1968)1339.
16 K.A. Johnston, Doctoral Thesis, University of Strathclyde, Glasgow, 1984; See also: K.A. Johnston and R.B. Duckworth in ref. 17, p. 65.
17 D. Simatos and J.L. Multon, Properties of water in foods, Martinus Nijhoff Publ., Dordrecht, 1985.
18 T. Roth and M. Loncin, in ref. 17, p. 331.
19 T.P. Labuza, Proc. 3rd Int. Conf. Food Sci. Technol., Washington, 1970, p. 618.
20 J.E.L. Corry, J. Appl. Bact., 44(1978)97.
21 L.R. Beuchat, J. Food Prot., 46(1983)135.
22 D.P. Bone, Food Prod. Dev., 3(1969)81.
23 R.B. Duckworth, CPCIA Seminar on Microbiology of low and intermediate moisture foods, Paris, 1972.
24 M. Karel, in ref. 17, p. 153.
25 C. van den Berg, F.S. Kaper, J.A.G. Weldring and I. Wolters, J. Fd. Technol., 10(1975)589.
26 S. Brunauer, L.S. Deming, W.E. Deming and E. Teller, J. Am. Chem. Soc., 62(1940)1723.
27 S. Brunauer, P.H. Emmett and E. Teller, J. Am. Chem. Soc. 60(1938)309.
28 C. van den Berg, Doctoral Thesis, Agricultural University Wageningen, 1981.
29 R. Heiss, Haltbarkeit und Sorptionsverhalten wasserarmer Lebensmittel, Springer Verlag Berlin, 1968.
30 H.A. Iglesias and J. Chirife, Handbook of food isotherms, Academic Press, New York, 1982.
31 W. Wolf, W.E.L. Spiess and G. Jung, Berichsheft Fachgemeinschaft Allgemeine Lufttechniek VDMA, Heft 18, Frankfurt/M, 1973.
32 R. Jowitt, F. Escher, B. Hallstrom, H.F.Th. Meffert, W.E.L. Spiess and G. Vos, Physical properties of foods, Applied Science Publ., London, 1983, Part 1.
33 W. Wolf, W.E.L. Spiess and G. Jung, in ref. 17, p. 661.
34 S. Gal, Die Methodik der WEasserdampf-Sorptionsmessungen, Springer Verlag, Berlin, 1967.
35 S. Gal, In R.B. Duckworth (ed), Water relations of foods, Academic Press, London, 1975, p. 139.
36 S. Gal, in L.R. Rockland and G.F. Stewart (eds), Water activity: influences on food quality, Academic Press, New York, 1981, p. 89.
37 J. Troller, J. Fd Prot. 46(1983)129.
38 O. Fennema and L.A. Berny, Proc. 4th Int. Conf. Food Sci. Technol., Madrid, 2(1974)27.
39 H. Bizot and J.L. Multon, Ann. Technol. Agric., 27(1978)441.
40 A.W. Boeke (ed), Het meten van luchtvochtigheid, P.C. Noordervliet, Zeist, The Netherlands, 1974.
41 K. Schurer, in ref. 17, p. 647.

42 T.D. McCune, K.W. Lang and M.P. Steinberg, J. Food Sci., 46(1981)1978.
43 A. Lenart and J.M. Flink, Lebensm.-Wiss.u-Techn., 16(1983)84.
44 A. Landrock and B. Proctor, Fd Technol., 5(1951)332.
45 J.B. Taylor and J.S. Rowlinson, Trans Faraday Soc., 51(1955)1186.
46 R.S. Norrish, J. Fd Technol., 1(1966)25.
47 L. Chuang and R.T. Toledo, J. Food Sci., 41(1976)922.
48 C. Ferro Fontan, J. Chirife and R. Boquet., J. Fd Technol., 16(1981)21, 553.
49 J. Chirife, C. Ferro Fontan and E.A. Benmergui, J. Fd Technol., 15(1980)59.
50 D.W. Grover, J. Soc. Chem. Ind., 66(1947)201.
51 R.W. Money and R. Born, J. Sci. Fd Agric., 2(1951)180.
52 K.D. Ross, Fd Technol., 29(1975)26.
53 C. Ferro Fontan, E.A. Benmergui and J. Chirife, J. Fd Technol., 15(1980)47.
54 T.T. Teng and C.C. Seow, J. Fd Technol., 16(1981)409.
55 C. Ferro Fontan and J. Chirife, J. Fd Technol., 16(1981)219.
56 R.A. Robinson and R.H. Stokes, Electrolyte solutions, 2nd rev. ed., Butterworths, London, 1970.
57 L.A. Bromley, A.I. Ch.E. Journal, 19(1973)313.
58 G.L. Kusik and H.P. Meissner, Ind. Eng. Chem., Proc. Res. Dev., 12(1973)112.
59 K.S. Pitzer and J.J. Kim, J. Am. Chem. Soc., 96(1974)5701.
60 P.J. Flory, Principles of polymer chemistry, Cornell University Press, Ithaca, 1953.
61 N.W. Taylor, J.E. Cluskey and F.R. Senti, J. Phys. Chem., 65(1961)1810.
62 C. van den Berg, in ref. 17, p. 119.
63 K.W. Lang and M.P. Steinberg, J. Food Sci., 46(1981)670.
64 I. Tomka, Doctoral dissertation, University of Berne, 1973.
65 S. Gal, I. Tomka and R. Signer, Chimia 30(1976)65.
66 C. van den Berg, in J.V. McLoughlin and B.M. McKenna (eds), Research in food science and nutrition, Boole, Dublin, 1984.
67 H. Bizot, in ref. 32, p. 43.
68 C.J. Lomauro, A.S. Bakshi and T.P. Labuza, Lebensm,-Wiss.u.-Technol., 18(1985)111, 118.
69 W. Schar and M. Ruegg, Lebensm.-Wiss.u.-Technol., 18(1985)225.
70 H. Weisser, in ref. 17, p. 95.
71 E.A. Guggenheim, Applications of statistical mechanics, Clarendon Press, Oxford, UK, 1966.
72 R.B. Anderson, J. Am. Chem. Soc., 68(1946)686.
73 J.H. de Boer, The dynamical character of adsorption, 2nd ed., Clarendon Press, Oxford, 1968.
74 H. Salwin and V. Slawson, Fd Technol., 13(1959)715.
75 L. Riedel, Chem. Mikrobiol. Technol. Lebensmittel, 5(1977)97.
76 C. van den Berg and H.A. Leniger, Misc. Papers Landbouwhogeschool Wageningen, 15(1978)231.

CONTROL OF LIPID OXIDATION IN DRIED FOODS

M. KAREL
Department of Applied Biological Sciences
Massachusetts Institute of Technology, Cambridge, MA (U.S.A.)

ABSTRACT

The major oxidative reactions in dried foods are due to peroxidation of lipids. The major products of oxidation of lipids are the peroxides of these lipids. The peroxides, however, are themselves highly reactive and may react with other food constituents, including in particular proteins and vitamins, or they decompose to secondary products, such as aldehydes, ketones, hydroxyacids and hydrocarbons, which are often volatile and may have strong odors and off-flavors. It is these smaller, volatile compounds that cause organoleptic rancidity of oxidized foods.

The oxidation reaction is a free radical reaction with an initiation step, a propagation step and a termination step. The rate is strongly affected by environmental and compositional factors. All of the steps of the reaction are affected by temperature and by water activity. In general at the water activities existing in dried foods the reaction is very rapid.

Temperature effect is not quite as great as in the case of other food reactions, and the oxidation can progress very rapidly even at moderate temperatures.

The most important factors affecting the reaction are those which control the rate of initiation of free radicals. The rate of oxidation increases with the rate of initiation, and the antioxidant concentration remaining in the system is also depleted at a rate that is equal to the rate of initiation. Therefore, if the rate of initiation is high then even the addition of antioxidants is only marginally effective in retarding oxidation.

The steps leading to control of undesirable oxidation may be summarized as follows:

1. Minimize oxygen concentration
2. Minimize rate of initiation by
    a. Minimizing initial oxidation level or contamination with oxidized components,
    b. Minimize pro-oxidant (metal) level,
    c. Maintain water activity at optimal level.
3. Choose effective antioxidants and assure their addition to sensitive lipids, <u>before</u> significant level of oxidation is reached.

INTRODUCTION

The major oxidative reactions in foods are due to peroxidation of lipids and of some water-soluble vitamins. Major products of oxidation of lipids, such as fatty acids, glycerides and phospholipids are the peroxides of these lipids. These peroxides can then react with the other food constituents, or they may decompose to secondary products, such as aldehydes, ketones, hydroxyacids and hydrocarbons, which are often volatile, and may have strong odor and off flavor. It is these smaller, volatile compounds that cause rancidity of foods. These secondary products as well as peroxides and lipid free radicals can react with proteins and vitamins causing losses in nutritional value and in solubility of food constituents.

The overall reaction involved in peroxidation in food and related systems is shown schematically below:

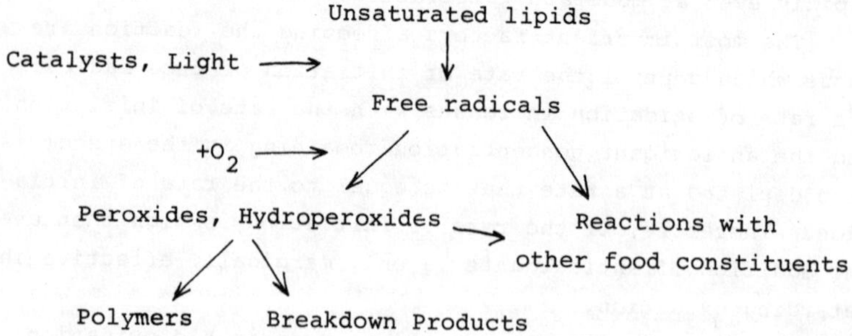

Mechanisms of peroxidation

Lipid oxidation in foods is associated almost exclusively with unsaturated fatty acids and is often autocatalytic, with oxidation products themselves catalyzing the reaction so that the rate increases with time. The reaction may also be catalyzed by metals or the enzyme lipoxidase. Both the catalyzed and autocatalytic reactions are initiated by decomposition of hydroperoxides to free radicals. In pure systems, monomolecular decomposition occurs up to at least 1-2% oxidation (molar basis), with the rate proportional to the square root of the extent of oxidation. During this period all oxygen absorbed forms hydroperoxides. Above 1% oxidation, bimolecular decomposition of hydroperoxides becomes controlling, the rate being proportional to the peroxide concentration.

The steps may be summarized as follows:
(RH refers to an unoxidized substrate, e.g. fatty acid, or fatty ester)

Initiation

$$RH \rightarrow R\cdot \quad \text{Requires powerful catalysts or radiant energy}$$

$$ROOH \rightarrow RO\cdot + \cdot OH$$

or

$$2ROOH \rightarrow RO\cdot + ROO\cdot + H_2O$$

The initiation by decomposition of hydroperoxides is by far the most important autocatalytic mechanism.

The decomposition of hydroperoxides to yield free radicals is also promoted very effectively by metal catalysts, such as copper or iron, as shown below, using as an example iron as catalyst.

$$ROOH + Fe^{++} \rightarrow RO\cdot + OH^- + Fe^{+++}$$

$$ROOH + Fe^{+++} \rightarrow ROO\cdot + H^+ + Fe^{++}$$

$$\text{Net: } 2\ ROOH \rightarrow RO\cdot + ROO\cdot + H_2O$$

An even more potent mechanism of peroxidation involves the enzyme lipoxygenase, which is much more effective than non-complexed metals. Various non-enzymic complexes of metals can have catalytic activities which either exceed, or are much lower than those of uncomplexed metals depending on the nature of the complex and on the composition of the food. Of particular importance in foods are heme-complexes, including various pigments present in both animal- and plant-derived products.

As mentioned, the autocatalytic (hydroperoxide catalyzed) reaction is considered to be predominant in absence of light and/or catalysts and is monomolecular at low concentrations of [ROOH] and bimolecular at high concentrations.

However, the concept that bimolecular decomposition of hydroperoxides is the predominant mode of initiation at relatively high hydroperoxide concentrations has been challenged by Hiatt and McCarrick (1975), who presented evidence that 3-hydroperoxy-2,3-dimethyl-1-butene decomposition in solvents involves addition of a peroxy radical to a hydroperoxide and proceeds as follows:

$$RO_2^{\cdot} + RO_2H \rightarrow RO_2R + \cdot OH$$

$$RO_2 \rightarrow 2RO\cdot \text{ (Manual**)}$$

This reaction has a concentration dependence similar to the bimolecular decomposition, but the mechanism is different. It is not known whether this mechanism is relevant in peroxidation of dry foods.

Propagation

Free radicals produced in the initiation step react rapidly with O_2, and the peroxides formed in this reaction propagate the chain by attacking an unoxidized substrate molecule to give back a free radical, and the major product - hydroperoxide.

$$R\cdot + O_2 \rightarrow RO_2\cdot$$

$$RO_2^{\cdot} + RH \rightarrow R\cdot + ROOH$$

Termination

The chain is terminated by reactions of free radicals to give nonradical products:

ROO· + ROO· → Non Radical Products

R· + R· → " " "

R· + ROO·· → " " "

An extremely important and useful route of termination is the reaction of free radicals with antioxidants (AH). Important antioxidants used in foods include the tocopherols, as well as the synthetic antioxidants butylated hydroxy anisol (BHA) butylated hydroxy toluene (BHT), propyl gallate (PG), tert-butylhydroquinone (TBHQ) and others. Many naturally occurring substances in plant materials, as well as some of the products of the Maillard reaction also possess antioxidant ability. The termination step (assuming R· is the dominant free radical species present) is

R· + AH → RH + A·

The free radical A· may be reconverted to AH by synergists such as ascorbic acid, as in the following sequence (Niki et al., 1984)

ROO· ⤵ ⤹ Vitamin E ⤴ ⤴ Vitamin C ·
ROOH ⤶ Vitamin E· ⤷ Vitamin C

The free radical A· may also be active as an antioxidant, as shown below

$$A\cdot + RO_2\cdot \rightarrow ROOA$$

The mechanisms of antioxidation are complex, and volumes have been written on the subject. Some recent references are Simic and Karel (1980) and Allen and Hamilton (1983).

Kinetics of oxidation

The presence of a variety of competing free radical reactions results in very complex reaction kinetics. However, these kinetics may be simplified very greatly by focusing on only one of the several potential indicators of oxidation used in assessing food quality listed in table 1, and by making simplifying assumptions concerning the course of the reaction. Under these circumstances, it is possible to write the simplified equations shown in figure 1.

TABLE 1

Indicators of oxidation used in assessing food quality

PRIMARY:
Peroxides, Hydroperoxides
Oxygen absorption
Diene conjugation
Loss of unsaturated fatty acids
SECONDARY:
Carbonyls (hexanal)
Malondialdehyde and related compounds (TBA)
Pentane
Reactions with other compounds (amino compounds)

R_O = RATE OF OXYGEN ABSORPTION

R_I = RATE OF INITIATION

NO ANTIOXIDANT, HIGH O_2 PRESSURE

$$R_O = K_1 (RH) (R_I)^{\frac{1}{2}}$$

NO ANTIOXIDANT, LOW O_2 PRESSURE

$$R_O = K_2 (O_2) (R_I)^{\frac{1}{2}}$$

ANTIOXIDANT, LOW O_2 PRESSURE

$$R_O = K_3 (O_2) (AH)^{-1} (R_I)$$

RATE OF LOSS OF ANTIOXIDANT = R_A

$$R_A = R_I$$

Fig. 1. Greatly simplified oxidation kinetics.

These equations show the critical importance of minimizing the rate of initiation. No matter what the other conditions are, the rate of oxidation increases with the rate of initiation, and the antioxidant concentration remaining in the system is also depleted at a rate that is equal to the rate of initiation. Therefore, if the rate of initiation is high, then even the addition of antioxidants is only marginally effective in retarding oxidation.

Effect of water activity

Water activity is a major factor affecting the rate of lipid oxidation in foods. It has long been recognized that very low water contents in fat-containing foods are conducive to rapid oxidation, and the literature documenting this phenomenon is voluminous (Karel and Yong, 1981).

Several hypotheses have been advanced to explain water's retarding effect on lipid oxidation. One factor that may be important is production of antioxidants through nonenzymatic browning. The effectiveness of products of nonenzymatic browning was demonstrated early by Griffith and Johnson (1957) and was one of the mechanisms hypothesized by Karel (1960). Several investigators in Japan and Europe have confirmed recently that intermediates in the complex set of reactions which we term nonenzymatic browning are effective antioxidants. Furthermore, intermediate moisture contents would maximize the concentration of these intermediates (Eichner, 1981). Since high water activity promotes browning, this may be one of the explanations for the observed effects.

Studies on purified model systems, containing no components capable of forming antioxidants through browning, however, also show retardation by increasing water content, and the explanation for these effects is probably based on the effect of water on initiation and termination steps of the chain reaction.

In purified systems, water interferes with the normal bimolecular decomposition of hydroperoxides formed at the lipid-water interface. In the presence of trace metals added to model systems of lyophilized emulsions, humidification at moderate levels retards oxidation because of hydration of metal ions. The reduction in rate depends on the type and hydration state of the metal salt added, as well as on water content. The monomolecular rate period is primarily affected by this mechanism, although bimolecular rates are also decreased (Karel, et al., 1967).

Relatively less is known about water activity's effect on termination, but it is to be expected that water would increase mobility and hence promote recombination of free radicals.

Water may also be expected to influence free-radical interactions between oxidizing lipids and other food components by influencing the concentrations of initiating radicals present, the degree of contact and mobility of reactants, and the relative importance of radical transfer versus recombination reactions. In particular, the reactions between peroxidizing lipids and proteins are strongly affected by water activity. Increasing water activity promotes the formation of Schiff base type of crosslinks between proteins, with bifunctional fragments of oxidized lipids incorporated as crosslinking agents. On the other hand at low water activities free radical crosslinks predominate, and the proteins crosslink without the incorporation of lipid fragments (Funes and Karel, 1981).

Retardation of peroxidation by increasing water content is reversed at high water activity. As water contents are increased above monolayer coverage in foods and in model systems, resistance to diffusion of solutes decreases and solubilization becomes significant. In systems containing chelating agents and antioxidants, high water contents allow solubilization of chelating agents that sequester metals as well as solubilization of antioxidants. This effect lowers the oxidation rate. However, at high water activities water may <u>accelerate</u> oxidation by solubilizing catalysts or by inducing swelling macromolecules such as proteins to expose additional catalytic sites.

A little-explored aspect is water's effect on food structure and on oxygen's ability to reach oxidation-susceptible food components. This aspect has been investigated thoroughly in recent years by Flink and coworkers (Karel and Flink, 1983). They studied the influence of physical structure of freeze-dried emulsified systems on the oxidation behavior of the lipid component. The partitioning of lipid into encapsulated and surface fractions was evaluated microscopically and chemically in emulsion systems of varying composition and concentrations, prepared under selected processing conditions and drying behavior. These studies demonstrated the influence of structure formation and lipid partitioning on subsequent oxidation behavior.

In systems in which there are both surface lipids, and lipid encapsulated within a carbohydrate, polysaccharide, or protein matrix the surface lipids oxidize readily when exposed to air. The encapsulated lipids, however, do not oxidize until the structure of the encapsulated matrix is modified or destroyed by adsorption of water. Prevention of this structure is of considerable importance in protecting unsaturated lipids from oxidation in dry preparation.

The most dramatic demonstration of the increased capability of various reactants to diffuse as water activity increases is provided in "glasses" formed by freeze-dried, or spray-dried, sugar-containing foods. When these glasses are humidified to higher water activities, they reach the "collapse" condition at which some mobility is possible. If crystallization is thermodynamically favored the crystallizable components are able to form crystals.

The same mobility which allows the glasses to be converted from an amorphous state to a mixture of crystals and amorphous components also allows the diffusion of various gases and vapors to encapsulated components within the glass. The importance of this resistance to diffusion in foods is that fat enrobed within such glasses (e.g., in candies, or in spray-dried milk) will not oxidize until the water activity of the matrix makes it sufficiently permeable to allow oxygen to penetrate. In this respect food glasses act like many plastics which also become more permeable when their plasticizer content increases. Table 2 shows how plasticizing a system based on sugar glasses facilitates diffusion of oxygen to droplets of linoleic acid entrapped in such glasses and allows rapid oxidation of this lipid (Geil Hausen and Flink, 1977).

TABLE 2

Oxygen absorbed after 90 hr by hexane-washed freeze-dried emulsions exposed to different relative humidities[a]

R.H. (%)	Oxygen absorbed (l/g) Linoleic acid (17.4%) entrapped in maltodextrin	Linoleic acid (20%) entrapped in maltose
0	0	0
43	0	190
75	0	200
93	100	280

Effect of oxygen concentration on oxidation of lipids in foods

When the effect of oxygen on oxidation of lipids in processing is considered, it is often simply a question if the total amount available for reaction with food components is, in fact, significant; and consideration must be given to the effect of concentration (or partial pressure) of oxygen on the rate.

The oxidation of unsaturated lipids is a free radical reaction with an initiation step, a propagation step and a termination step. The rate is strongly affected by environmental and compositional factors, and these are summarized in Table 3. As indicated in this table all of the steps of the reaction are affected by temperature and by water activity, which were discussed previously.

TABLE 3

Factors controlling rate of oxidation

INITIATION:	
	Hydroperoxide concentration
	Presence of catalysts
	Electromagnetic radiation (includes light)
	Chemical initiators (includes some oxidation products)
PROPAGATION:	
	Degree of unsaturation
	Local oxygen concentration
TERMINATION AND BRANCHING:	
	Antioxidant concentration
TEMPERATURE AND WATER ACTIVITY ARE FACTORS IN ALL OF ABOVE STEPS.	

These equations show how important it is to minimize the rate of initiation of free radicals. No matter what the other conditions are, the rate of oxidation increases with the rate of initiation, and the antioxidant concentration remaining in the system is also depleted at a rate that is equal to the rate of initiation. Therefore, if the rate of initiation is high, then even the addition of antioxidants is only marginally effective in retarding oxidation.

A general form often found useful in correlating reaction rate with oxygen concentration is given below (Karel, 1960):

$$R = \frac{[O_2]}{k_1 + k_2[O_2]}$$

where R = reaction rate

$[O_2]$ = oxygen concentration

k_1, k_2 = constants

Note that when $k_2[O_2]$ is much smaller than k_1, the rate is linear with oxygen concentration, and when $k_2[O_2]$ is much larger than k_1, the rate becomes independent of concentration. Most recently the validity of this equation was tested in dehydrated systems with good results by Kacyn et al. (1983).

The effect of oxygen concentration present at any time within the oxidizing system on the rate of oxidation is shown schematically in Fig. 2. It should be evident that in order to avoid oxidation, very low oxygen concentrations are necessary. This theoretical conclusion was confirmed by work conducted at the U.S. Army Natick R&D Labs on military rations (Tuomy et al., 1968a, b; 1969; 1970). It should be noted, however, that this relation applies to concentrations of oxygen in the locations in which the reactions take place. When diffusional limitations exist, or when oxygen is desorbed slowly from a freeze-dried powder, the local oxygen concentrations may be different from those that would be expected if the headspace and the freeze-dried ration component were in instant, and continuously maintained equilibrium (Bishov et al., 1971; Quast and Karel, 1971; Teixeira-Neto et al., 1981).

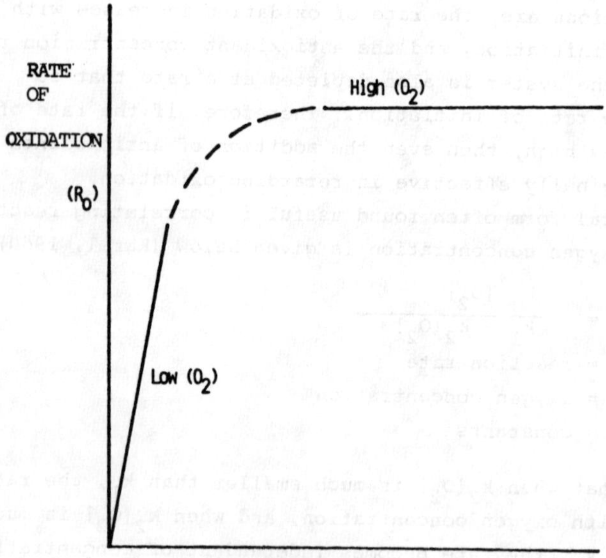

Fig. 2.

Control of oxidation in dry foods

The practical steps one can take based on the above theoretical considerations are listed in Table 4 (Karel, 1984). It should be obvious that the minimizing of the oxygen concentration would require effective removal of oxygen from within the dry food as well as from headspace. We have stressed above the need to minimize free radical initiation. In addition to the pro-oxidant factors listed in Table 3, it should be also pointed out that exposure to light and other radiations will catalyze initiations and should therefore be avoided. Another group of initiators often neglected is due to volatile decomposition products of oxidation. El-Magoli et al. (1980) have shown that oxidation of foods could be accelerated not only by direct incorporation of pre-oxidized lipids, but even by exposure to volatile breakdown products from oxidizing lipids. Therefore, even the presence in the same container of an already oxidized food will tend to speed up the oxidation of an unoxidized food. In choosing antioxidants it is also important to add them in a manner which assures good contact with the lipids which it is to protect, and also that the antioxidant be added before the oxidation of the food has progressed very far.

TABLE 4
Principles of control of oxidation in dry foods which follow from kinetic principles

1. Minimize oxygen concentration
2. Minimize rate of initiation by
    a. Minimizing initial oxidation level or contamination with oxidized components,
    b. Minimize pro-oxidant (metal) level,
    c. Maintain water activity at optimal level.
3. Choose effective antioxidants and assure their addition to sensitive lipids, <u>before</u> significant level of oxidation is reached.

Another important consideration is the exclusion from the dehydrated food of oxidizing agents during the dehydration process. Recent work by Tsai and Hudson (1985) has confirmed that cholesterol oxidation products, including potentially dangerous cholesterol α-oxides can be detected in commercial dry yolk, and that the presence of these oxides is strongly correlated with the use of direct gas fired driers. It is strongly suspected that the oxidation of the cholesterol is linked to the presence of nitrogen oxides.

In a direct gas fired system, liquid egg was dried by the flue gas which contained the combustion products of natural gas with an empirical chemical formula $C_1H_{3.8}$. Complete combustion of natural gas produces water, carbon dioxide and minor amounts of carbon monoxide (less than 10 ppm), nitrogen dioxide and nitric oxide. The last two compounds are usually referred collectively as nitrogen oxides or NO_x. Concentrations of NO_x in flue gas vary from 80-160 ppm in home hot water boilers and commercial cooking ovens to several hundred parts per million in large utility boilers. Nitrogen oxides are known stable free radicals. They may serve as initiators of cholesterol autoxidation (Smith, 1981).

ACKNOWLEDGEMENT

This work was supported in part by grants no. CPE-8104582 and CBT-8420689 from the Engineering Division of the National Science Foundation.

REFERENCES

1 J.C. Allen and R.J. Hamilton, Rancidity in Foods, Appl. Sci. Publ., London, 1983.
2 S.J. Bishov, A.S. Henick, J.M. Giffee, I.T. Nii, P.A. Prell, and M. Wolf, J. Food Sci., 36(1971)532.
3 K. Eichner, Antioxidative effect of Maillard reaction intermediates. In: C. Eriksson (Editor) Maillard Reactions in Food. Pergamon Press, Oxford, p. 441, 1981.
4 S. El-Magoli, M. Karel, and S. Yong, J. Food Biochem. 3(1980)111.
5 J. Funes and M. Karel, Lipids, 16(1981)347.
6 F. Geil-Hansen and J.M. Flink, J. Food Sci., 42(1977) 1049.
7 T. Griffith and J. Johnson, Cereal Chem., 34(1957)159.
8 R. Hiatt and T. McCarrick, J. Am. Chem. Soc., 97(1975), 5234.
9 L.J. Kacyn, I. Saguy and M. Karel, J. Food Process. Preserv., 7(1983)193.
10 M. Karel, Effects of water and oxygen on chemical reactions, Ph.D. Thesis (1960), M.I.T., Cambridge, MA.
11 M. Karel, T.P. Labuza and J.F. Maloney, Cryobiology, 3(1967)1288.
12 M. Karel and J.M. Flink, Some recent developments in food dehydration research. In: A.S. Mujumdar (Editor), Advances in Drying, Vol. 2, Hemisphere Publ. Corp., Washington, D.C., 103-154, 1983.
13 M. Karel and S. Yong, Autoxidation-initiated reactions in foods. In: L.B. Rockland and G.F. Stewart (Editors) Water Activity: Influences on Food Quality. Academic Press, New York, 1981.
14 M. Karel, Activities Report, 36(1984)1: 50.
15 E. Niki, T. Saito, A. Kawakami and Y. Kamiya, J. Biol. Chem., 259(1984)4177.
16 D. Quast and M. Karel, J. Food Technol., 6(1971)95.
17 M. Simic and M. Karel (eds.), Autoxidation in food and biological systems, Plenum Press, New York, 1980.
18 L.L. Smith, Cholesterol autoxidation, Plenum Press, New York, 1981.
19 R. Teixeira-Neto, M. Karel, I. Saguy, and S. Mizrahi, J. Food Sci., 46(1981)665.
20 L.S. Tsai and C.H. Hudson, J. Food Sci., 50(1985)229.
21 J. Tuomy and L. Hinnedgardt, Technical Report 68-65-FL, U.S. Army Natick R&D Laboratories, Natick, MA, 1968a.
22 J. Tuomy and L. Hinnegardt, Technical Report 69-54-FL, U.S. Army R&D Laboratories, Natick, MA, 1968b.
23 J. Tuomy, L. Hinnegardt, and R. Helmer, J. Agric. Food Chem., 17(1969)1360.
24 J. Tuomy, L. Hinnegardt, and R. Helmer, J. Agric. Food Chem., 18(1970)899.

MAILLARD BROWNING REACTIONS IN DRIED FOODS

H.E. NURSTEN
Department of Food Science, The University, Whiteknights, Reading (U.K.)

ABSTRACT

The Maillard reaction is still best summarised by Hodge's scheme (Fig. 1; ref. 1). Its complexities are outlined and its symptoms listed, together with their significance for the Food Industry. Recent work on determining the nature of the chromophores produced is reviewed before considering the effect of a_w on Maillard browning and the bearing of the two major pathways of the reaction (Fig. 4). Progress has been made recently, using ^{13}C-NMR in particular, in understanding the structure of melanoidins, but the methods for controlling the reaction remain largely traditional.

INTRODUCTION

The Maillard reaction is still best summarised by Hodge's scheme (see over).

Four points should be noted:

1. Sugars on their own, provided they are exposed to higher temperatures, will undergo similar reactions by themselves (caramelisation), but then there can be no interaction with amino compounds and no nitrogen-containing compounds can result.

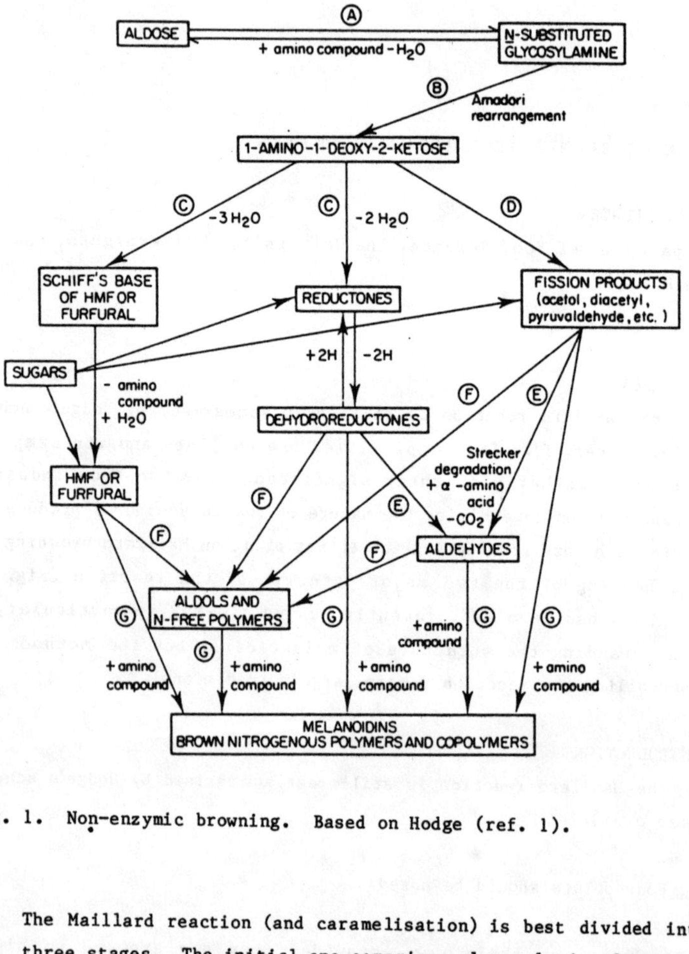

Fig. 1. Non-enzymic browning. Based on Hodge (ref. 1).

2. The Maillard reaction (and caramelisation) is best divided into three stages. The initial one comprises glycosylamine formation (A) and subsequent rearrangement (B). The intermediate stage comprises dehydration (C), either by loss of three molecules of water to furfurals or by loss of two to reductones; fission (D), mainly by dealdolisation; and Strecker degradation (E), the interaction of amino acids and dicarbonyl compounds, which may be either dehydroreductones or dehydration or fission products. The final stage consists of the conversion of carbonyl

compounds, be they furfurals, fission products, dehydroreductones, or Strecker aldehydes, into high molecular weight products, either nitrogen-free (F) or nitrogenous (G), when amines are available.

3. Ketoses undergo similar reaction, reaction B then being termed the Heyns rearrangement.
4. Some of the reactions are reversible, notably the first step A and the dehydrogenation of reductones.

There are 12 symptoms of the Maillard reaction:

1. Production of colour or discoloration.
2. Production of flavour or off-flavour.
3. Production of water.
4. Production of carbon dioxide.
5. Lowering of pH.
6. Increasing reducing power.
7. Decreasing solubility.
8. Loss of Vitamin C activity.
9. Loss of biological value of protein.
10. Chelation of metals.
11. Production of toxicity.
12. Production of fluorescence.

Their significance to the Food Industry is manifold:

1. Changes in appearance, mainly in colour. Browning may be desirable, as in coffee and bread crust, or undesirable, as in glucose syrup and in many intermediate moisture products. Decreased solubility may have implications for appearance, as in lessened solubility of skim milk powder.
2. Production of flavour or off-flavour. Odours and off-odours are due to volatile products, e.g., fission products and Strecker aldehydes. Substances tasting sweet or bitter may be involved, as well as effects on texture.
3. Reduction in nutritional value by involvement of ascorbic acid

(a reductone) and of lysine (free or bound), an essential and often limiting amino acid. Metal-chelating properties may also be significant.

4. Toxicity through the possible formation of imidazoles and of N-nitroso derivatives, e.g., of Amadori compounds. The intrinsic toxicity of Maillard products and intermediates is also under study, but is made difficult by their anti-nutritional properties.
5. Antioxidant properties. These are thought to be due to the reductones formed, but chelation of heavy metals may also be involved.
6. Increased a_w because of production of water.

At this symposium the concern has primarily been with the concentration and drying of foods, i.e., with foods as their a_w has been reduced. The focus here, therefore, needs to be on how lowering a_w affects the Maillard reaction.

CHROMOPHORES PRODUCED BY THE MAILLARD REACTION

As the Maillard reaction is complex, the effects on it will be ramified. What effects are of interest? All the 12 symptoms listed above, except perhaps the last, are of significance to the Food Industry and elsewhere, as for those concerned with blood, particularly that of diabetics, intravenous feeds, cataract, ageing, leprosy, humic substances of the soil, the Paper Industry, and cellulosic textiles.

Colour production is the most obvious symptom and has often been used as the yardstick. Colour is not produced in the initial stage of the reaction, only little is produced in the intermediate stage, and most of it results in the final stage. The Maillard reaction could be considered in a similar way in relation to each of the other symptoms. In each case, a profile modified to a greater or lesser extent would emerge; for example, much is known about lysine ceasing to be available as the Maillard reaction progresses through the initial stage only. However, here the focus will be on colour.

Surprisingly little is known of the structure of any chromophores formed. Severin and Krönig (ref. 2) isolated the yellow compound 1, X = O, from mixtures produced by heating xylose with isopropylammonium acetate and found that 1, X = O, was also formed when xylose was replaced by arabinose or isopropylammonium acetate by amino acids, such as lysine or glycine. Ledl and Severin (ref. 3) state that pentoses and methylammonium acetate lead to 1, X = NCH_3, yellow, in addition, and that xylose and glycine similarly give the orange compound 1, X = NCH_2COOH.

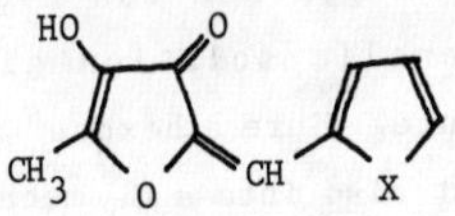

(1) X = O, NCH_3, NCH_2COOH

The latter authors also found that the methyl group of compounds such as 1 is sufficiently reactive to condense with carbonyls to form compounds such as 2 (R,X = H,O, deep orange; CH_3,O; or H,NCH_3).

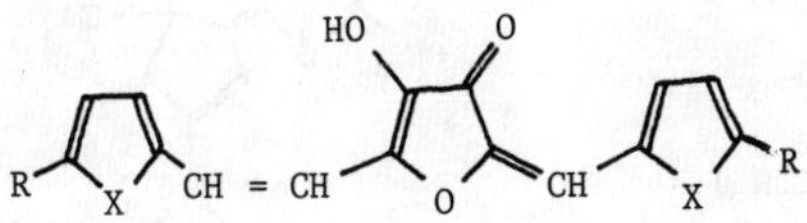

(2) R = H, CH_3; X = O, NCH_3

Hexoses can also undergo similar reactions. Thus, Ledl and Severin (ref. 4) heated the Amadori compound from glucose and piperidine with furfural and obtained 3. From the reaction of glucose and methylammonium acetate, Ledl <u>et al.</u> (ref. 5) isolated 4-6.

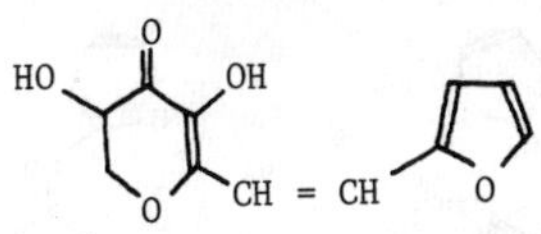

(3)

4 was obtained in yellow-orange crystals, 6, R = H or CH_3, the structure of which is not certain, is yellow, and 5 has λ_{max} = 406 nm.

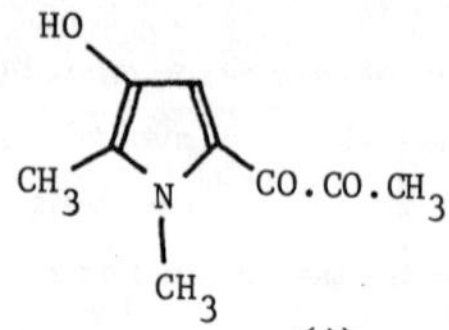
(4)

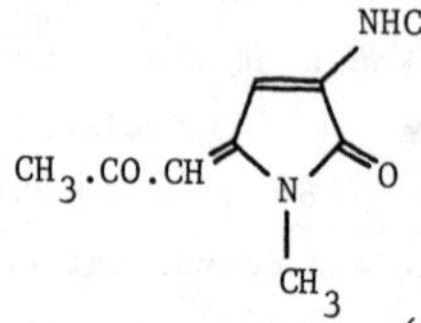
(5)

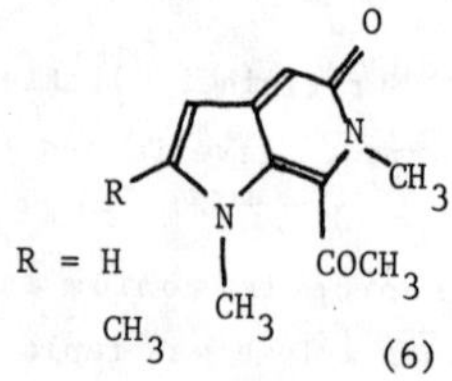
(6)

The 8-C chain in 4 is a reminder that the 6-C chain of glucose does undergo fission, followed by recombination; 5 may be formed in a parallel manner, but the pathway to 6 is not yet clear.

In methanolic solution, xylose and diethylammonium acetate gave the orange 7, which shows that a pentose can cyclise to a cyclopentane. Similarly, in ethanolic solution glucose and piperidine acetate gave 8, also orange. Here a hexose is not only converted to a methylcyclopentane, but also into a C_5 methylfuranone, <u>i.e.</u>, fission again, the loss of formaldehyde by a retroaldol condensation having probably taken place (ref. 5).

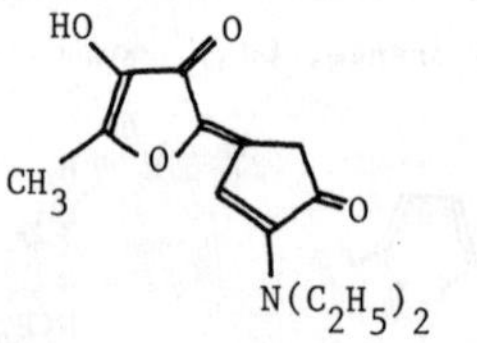
(7)

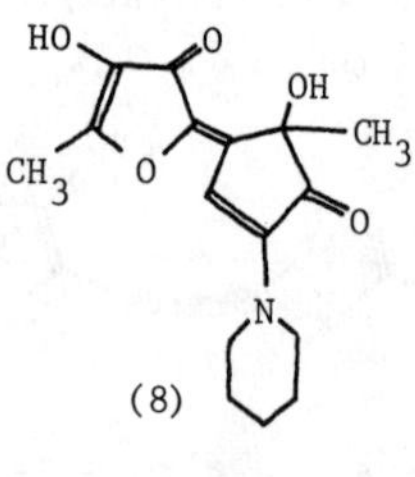
(8)

Ledl <u>et al</u>. (ref. 6) examined the reaction of xylose and glycine in methanolic solution and isolated 9 and 10, both yellow. Glucose in place of xylose similarly yields 11.

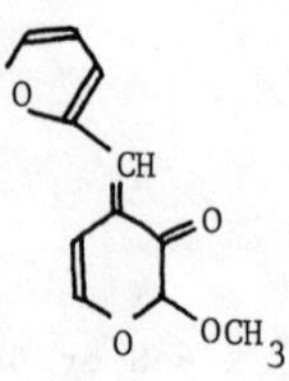
(9)

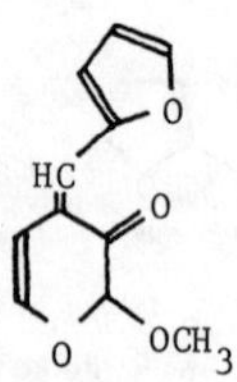
(10)

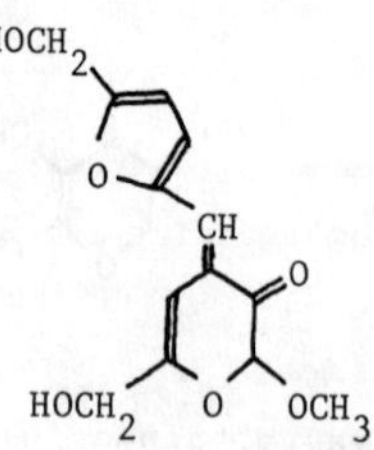
(11)

Even more complex coloured compounds, with up to four such rings in them, have been isolated from the reaction products from xylose and glycine by painstaking chromatographic separations (ref. 7).

THE EFFECT OF a_w ON MAILLARD BROWNING

In spite of the work outlined above, the position has still not been reached where browning can be followed by determining a specific compound or chromophore. The spectra of browning products are almost invariably featureless, with absorption tapering away gently into the visible region. Most frequently, absorbance readings are taken at 420 nm as an indication of browning.

Browning assessed in this way varies with a_w, usually showing a maximum at intermediate values of a_w of 0.5-0.8. The literature is summarised in Table 1. Neglecting time and temperature for the

TABLE 1
Water activity for maximum rate of browning

Food	a_w	Reference
Apricots (under nitrogen)	0.25-0.45	*via* 8
Whey powder	*ca* 0.44	10
Pork bites	0.52	*via* 9
Dried meat	0.57	*via* 8
Pork sausage	0.62	*via* 9
Dried milk	0.68	*via* 8
Pea soup mix	0.69	*via* 9
Potatoes	0.73	*via* 8
Corn chowder mix	0.79	*via* 9
Egg noodles	$\not< 0.75$	11

moment, the concentration and drying of foods runs into most trouble from browning as the food passes through these intermediate values of a_w. As regards shelf-life, it is again foods with such intermediate values of a_w that are most liable to deterioration from non-enzymic browning.

The inter-relations of some of these factors are well illustrated by experiments on air drying of carrot cubes by Eichner and Wolf (Fig. 2; ref. 12).

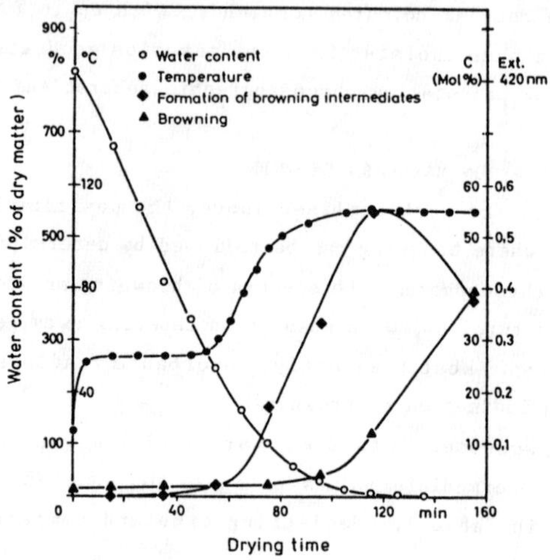

Fig. 2. Formation of browning intermediates and browning during air drying of carrot cubes (air temperature: 110°C). Reproduced by permission from ref. 12.

The following 4 stages are worth noting:

1. As the temperature of the cubes comes up, there sets in the constant drying rate, which lasts until diffusion of water from the interior of the cubes begins to be limiting.
2. Water continues to be lost, but the rate of loss begins to fall off and the temperature rises quite steeply.
3. The Maillard reaction is initiated, as assessed by the formation of some Amadori compounds by means of ion-exchange amino acid analysis of aqueous extracts without prior hydrolysis.
4. The formation of the Amadori compounds goes through a maximum as browning itself takes off, indicating that the colourless Amadori compounds are being converted into coloured products.

If the drying process had been stopped at about 120 min, the cubes would have exhibited very little browning, but considerable amounts of Amadori products would be present. These are very significant, as can be seen from Fig. 3.

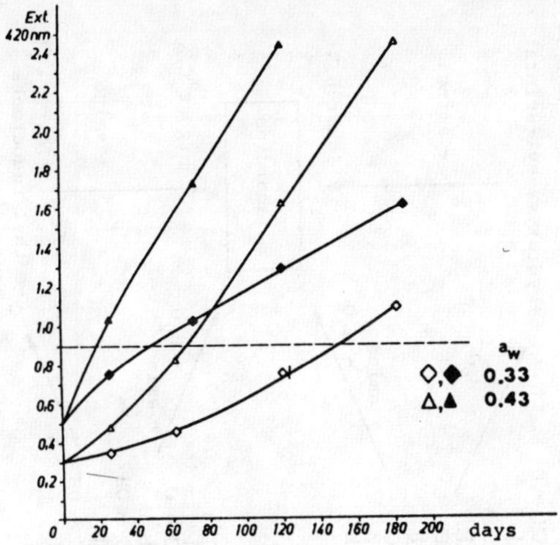

Fig. 3. The influence of heat treatment on the shelf-life of freeze-dried tomato powder stored at 23°C and different water activities (Ext. $_{420\ nm}$ = 0.9 corresponds to the limit of acceptability). ◆ ▲ preheated at 40°C and a water activity of 0.11 (about 30% of glutamic acid having reacted to fructosylglutamic acid); ◇ △ unheated. Reproduced by permission from ref. 12.

Here the browning of freeze-dried tomato powder, containing considerable amounts of Amadori products because of prior exposure to mild heating, is compared with unexposed powder. Shelf-life at 23°C is greatly curtailed, browning starting immediately at the highest rate.

It is worth looking at the reactions that lie between the Amadori compounds and the melanoidins in somewhat more detail (Fig. 4).

Several points need to be made:

1. Most of the reactions are equilibria, but irreversibility starts with the formation of the Amadori compound from the N-substituted glycosylamine and there appear to be irreversible steps in each of the two major pathways, that based on 1,2-enolisation and that on 2,3-enolisation.

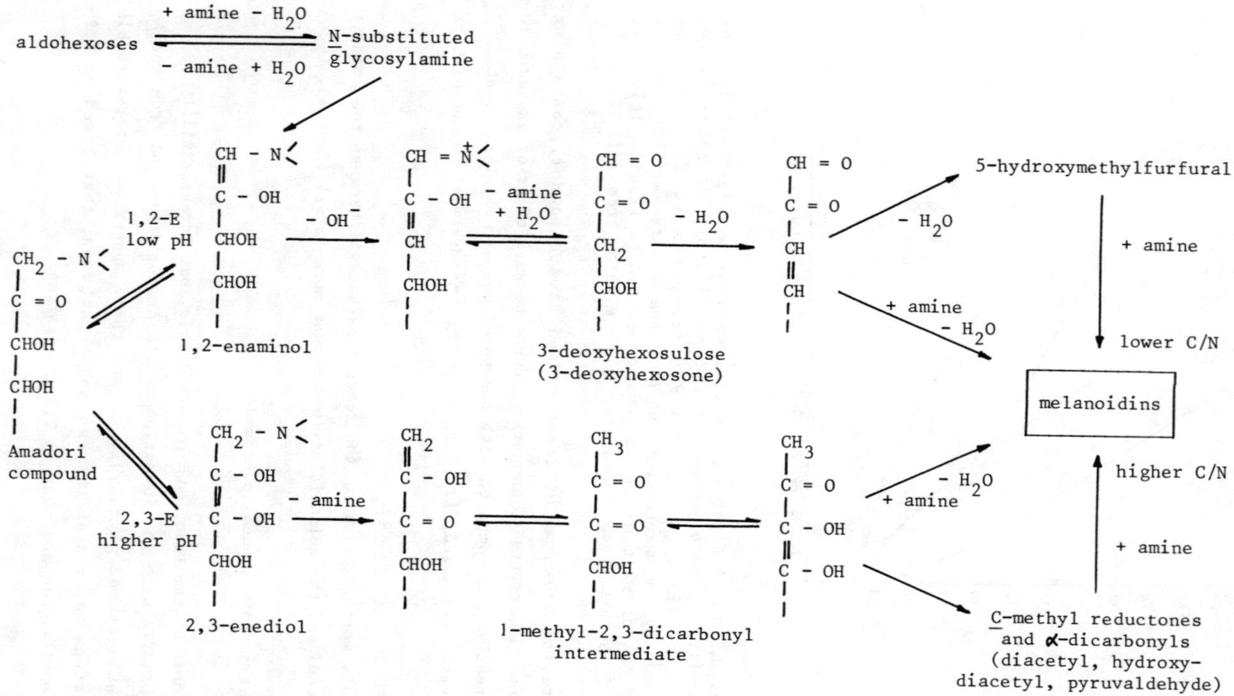

Fig. 4. Maillard reaction: two major pathways from Amadori compounds to melanoidins. Based on Hodge (ref. 13).

2. pH is a major determinant of the extent to which each pathway is followed; low pH favours 1,2-enolisation, high pH 2,3.
3. The effect of pH will be modified by the basicity of the amine, less basic amines being less susceptible to protonation and thus favouring 2,3-enolisation.
4. Each pathway leads to an -dicarbonyl compound: a 3-deoxyhexosulose and a 1-methyl-2,3-dicarbonyl compound, respectively.
5. Each pathway leads to furan derivatives: furfurals and methylhydroxyfuranones, respectively.
6. Of the coloured compounds given above, 1-3 and 9-11 are composed of moieties obtained by *both* 1,2- and 2,3-enolisation, *i.e.*, requiring reasonably subtle control for their formation.
7. Water is lost in Maillard browning, 3 mol in the formation of furfurals and 2 mol in that of methylhydroxyfuranones. Excess of water, *i.e.*, high a_w, thus tends to hinder the reaction by mass action as well as by diluting the reactants.

In contrast, at low a_w, even though the concentration of the reactants increases, they begin to lose mobility, and so it is not surprising that the Maillard reaction possesses an a_w for maximum reactivity. However, the presence of other substances can have an effect. Humectants, such as glycerol, move the maximum to lower values of a_w, for example, from a_w 0.7 to 0.5, to below the a_w of many intermediate moisture foods (ref. 9).

MELANOIDINS

Melanoidins are the end products of the Maillard reaction, but it is by no means clear what they are. They are considered to be both of high molecular weight and brown. Clearly they will differ depending on the reactants. Thus in reaction with glycine, glucose gives melanoidins high in aliphatic hydroxy groups, furfural ones high in ether oxygen, and pyruvaldehyde ones high in enolic hydroxy groups, but low in ether oxygen.

Very recent work by Benzing-Purdie *et al.* (ref. 14) bears on this (see Table 2).

TABLE 2

Composition of retentate of melanoidins from xylose and glycine[a]

Reaction conditions Temp., ^oC	Time	Yield, %[b] Retentate	Diffusate	Microanalytical data, % C	H	N
22	9 months	4.2	--	50.3	5.3	8.0
68	6 weeks	39.6	30.0	57.0	6.1	7.3
100	38 h	30.7	15.8	57.8	5.4	6.7

[a] molar ratio, 1:1
[b] based on total reactants; membrane cut-off, 12,000 daltons

The following points can usefully be made:

1. In relation to forming high molecular weight material, temperature seems to be much more important than time. The proportion of polymeric material in the product increased progressively with temperature.
2. Loss of material, presumably mainly as H_2O and CO_2, is much greater at 100° (54) than at 68° (30%).
3. The composition of the retentate formed at 22° corresponds quite closely to the loss of 3 mol H_2O: C, 49.1; H, 5.3; N, 8.2%, *i.e.*, all the nitrogen appears to be retained. Subsequently, nitrogen is lost, presumably as volatiles.

^{13}C-NMR indicated that all these melanoidins contained methyl groups, but the low-temperature one provided almost no evidence of aromatic or heteroaromatic carbon, carbon bearing oxygen or nitrogen in aromatic rings or carbon alpha to nitrogen, and contained less unsaturation and fewer carbonyl groups than the other two. The major increase with temperature in aromaticity without much loss of carbonyl groups was confirmed by infrared spectra and by ^{13}C-NMR of the diffusates. Whereas the diffusate from reaction at 68° showed the 12 resonances of the Amadori compound from 2 mol xylose and 1 mol glycine, as well as the presence of glycine and a little xylose, the diffusate from reaction at 100° was much more complex, with much greater aromatic character, and indications of azomethine, acid, anhydride, ester, and/or amide carbons.

Feather and Nelson (ref. 15) have also carried out some interesting experiments, using glucose and glycine (molar ratio, 1:1; at 1.0:0.1, yield and colour development were greatly reduced) in phthalate buffer, pH 3.5, under reflux. Fructose darkened more rapidly than glucose, but gave only half the yield. The polymers (membrane cut-off, 16,500 daltons) showed little electronic absorption. Their nitrogen contents were: glucose, 6.2; fructose, 6.9; hydroxymethylfurfural, 3.5%, elementary analysis of the first two suggesting loss of 3 mol H_2O (*cf.* above). Glucose and methionine led to N, 3.3; S, 4.7%, *i.e.*, suggesting incorporation of the amino acid as an entity. Here, ^{13}C-NMR was used in conjunction with glucose-1-^{13}C, glycine-1-^{13}C, and glycine-2-^{13}C. The diffusates contained as major components the sugar, amino acid, and the Amadori compound. The retentate from glycine-1-^{13}C showed signals corresponding to carbonyl carbon, so not all the carboxy groups can be lost as CO_2. Glycine-2-^{13}C duly led to signals consistent with substituted methyl groups and glucose-1-^{13}C did the same, *i.e.*, the carbonyl group at C-1 has been modified in conformity with Fig. 4 (confirmed by Kato *et al.*, ref. 16). Some polymer resonances found correspond to those of Amadori compounds, yet the loss of water is greater than at that stage of the Maillard reaction. Experiments with fructose and glycine-1- and -2-^{13}C demonstrate some structural differences from the corresponding glucose reaction products.

Similar results have been obtained even more recently by Feather and Huang (ref. 17), using *L*-alanine-1- and -2-^{13}C. The results suggest that C-1 is incorporated into the polymers as a carboxyl carbon and C-2 as a substituted amino carbon, C-1 of glucose becoming a substituted methyl group. Kato *et al.* (ref. 16) have shown that the mean molecular weight of the polymers from the glucose - glycine reaction is reduced to two thirds with alkaline peroxide and to less than half by ozonolysis. The polymers contain many acetyl groups and almost 6% releasable glycine.

CONTROL OF THE MAILLARD REACTION

There are five main methods:

1. Temperature control.
2. Sulphur dioxide.
3. pH.
4. a_w.
5. Composition.

Temperature control

Maillard browning has a relatively high temperature coefficient, the Q_{10} lying usually in the range 3-6 (ref. 9). Refrigeration is thus a good means of discouraging the reaction and increasing the temperature of encouraging it (cf. Table 2).

Sulphur dioxide

Sulphur dioxide is an important method of controlling not only Maillard browning, but enzymic browning and micro-organisms as well. the mechanism does not seem to depend on the formation of aldehyde bisulphite compounds, but rather on the addition of bisulphite across double bonds, particularly those of α,β- unsaturated carbonyl compounds, such as the 3,4-dideoxyhexosulos-3-ene (see Fig. 4). Even that is not the whole story (ref. 18).

pH

In general, lowering the pH reduces the degree of browning obtained, since the browning potential of 1,2-enolisation is less than of 2,3-enolisation (see Fig. 4). Very low pH in presence of ascorbic acid can lead to non-enzymic browning of a specific kind.

a_w

Reduced browning is to be found at high and low values of a_w, but, as explained above, many factors affect the position of maximum browning.

Composition

The Maillard reaction can be said to be characterised by the acceleration of the caramelisation of sugars through the presence of amines, usually amino acids, peptides, or proteins. In carbohydrate foods it can therefore be controlled by means of the amine and, conversely, in protein foods, by means of the reducing sugar. In the former case, concentration of soluble protein is more significant than total protein. Reactivity differs between amino acids, and lysine with its ε-amino group has a special place. Histidine and arginine have particular significance in relation to antioxidant properties (ref. 19). The subtleties of peptide composition, particularly in relation to flavour production, are beginning to be appreciated. As regards sugars, pentoses are much more reactive than hexoses and aldoses differ from ketoses (*cf.* above). Recently, Kato *et al.* (ref. 20) found that a mixture of ovalbumin with galactose stored 4 days at 50° and 65% R.H. browned about twice as much as mixtures with glucose or mannose, solubility also being greatly reduced. Comparisons between mono- and disaccharides are also beginning to be made. The role of third and further components is also coming to the fore (see above).

SUMMARY

From the above, in spite of the severe compression necessary, the great complexity of the Maillard reaction is abundantly illustrated. Using a range of modern methods, progress is currently being made in unravelling particularly the nature of the chromophores formed and of the reaction mechanisms operating at the different stages of the reaction. Such work needs to be related back to the composition of the reactant mix and the range of relevant reaction conditions, as well as forwards to the profiles produced in terms of each of the symptoms of practical concern. Much remains to be done, but the possibilites of further, rapid advances now seem so great, that there is great eagerness to thrust forward.

REFERENCES

1. J.E. Hodge, J. agric. Fd. Chem., 1 (1953) 928-943.

2. T. Severin and U. Krönig, Chem. Mikrobiol. Technol. Lebensm., 1 (1972) 156-157.
3. F. Ledl and T. Severin, Z. Lebensm.-Unters. -Forsch., 167 (1978) 410-413.
4. F. Ledl and T. Severin, Z. Lebensm.-Unters. -Forsch., 175 (1982) 262-265.
5. F. Ledl, U. Krönig, T. Severin and H. Lotter, Z. Lebensm.-Unters. -Forsch., 177 (1983) 267-270.
6. F. Ledl, J. Hiebl and T. Severin, Z. Lebensm.-Unters.-Forsch., 177 (1983) 353-355.
7. H.E. Nursten and R. O'Reilly, Fd Chem., in press.
8. K. Eichner, in R.B. Duckworth (Ed.), Water Relations of Foods, Academic Press, London, 1975, pp. 417-434.
9. T.P. Labuza and M. Saltmarch, in L.B. Rockland and G.F. Stewart (Eds), Water Activity: Influences on Food Quality, Academic Press, New York, 1981, pp. 567-603.
10. T.P. Labuza and M. Saltmarch, J. Fd. Sci., 47 (1982) 92-96, 113.
11. T.P. Labuza, H. Bohnsack and M.N. Kim, Cereal Chem., 59 (1982) 142-147.
12. K. Eichner and M. Ciner-Doruk, in L.B. Rockland and G.F. Stewart (Eds), Water Activity: Influences on Food Quality, Academic Press, New York, 1981, pp. 567-603.
13. J.E. Hodge, in H.W. Schultz, E.A. Day and L.M. Libbey (Eds), The Chemistry and Physiology of Flavors, AVI Publishing Co., Westport, Conn., 1967, p. 465-491.
14. L.M. Benzing-Purdie, J.A. Ripmeester and C.I. Ratcliffe, J. agric. Fd Chem., 33 (1985) 31-33.
15. M.S. Feather and D. Nelson, J. agric. Fd Chem., 32 (1984) 1428-1432.
16. H. Kato, S.B. Kim and F. Hayase, Amino-Carbonyl Reaction in Food and Biological Systems, (1985) Abstract C-26.
17. M.S. Feather and R.D. Huang, Amino-Carbonyl Reaction in Food and Biological Systems, (1985) Abstract C-22.
18. B.L. Wedzicha, Chemistry of Sulphur Dioxide in Foods, Elsevier Applied Science, 1984.
19. H. Lingnert and G. Hall, Amino-Carbonyl Reaction in Food and Biological Systems, (1985) Abstract T-39.
20. Y. Kato, T. Matsuda, N. Kato, K. Watanabe and R. Nakamura, Amino-Carbonyl Reaction in Food and Biological Systems, (1985) Abstract C-13.

LOSS OF NUTRITIVE VALUE ON DRYING

H.F. ERBERSDOBLER
Institut für Humanernährung und Lebensmittelkunde der Universität Kiel
Düsternbrooker Weg 17-19, D2300 Kiel 1, (F.R.G.)

ABSTRACT

The influence of drying on the nutritive value of food depends on the nutrient involved as well as on the drying technique used. In freshly dried foods, vitamin losses are generally less than natural variations in vitamin content, post-harvest losses before processing and normal losses during home cooking. Model experiments show that damage to protein quality can be divided into (a) total destruction of amino acids, (b) derivatisation of some individual amino acids and (c) cross-linking reactions between certain amino acids. The total destruction which occurs at higher temperatures is easily measured by amino acid analysis. Derivatisation of lysine in the Maillard reaction decreases its availability, and can be detected using the furosine method, and several examples are given. (The Maillard reaction is discussed in detail in another paper in this symposium). Crosslinking reactions can also reduce the availability of amino acids as is the case with lysinoalanine. Digestibility is also damaged if extensive crosslinking takes place.

The damage caused during drying of plant foods is demonstrated using the example of maize, for which the lysine loss was measured and animal feeding experiments conducted. It is shown that the main factors influencing the protein quality are temperature and time of heating, air velocity and thickness of the layer of corn in the drier.

A table showing lysine losses during the drying of some commercial foods of plant and animal origin is presented. Investigations of drying of milk are reported, in which substantial losses of some vitamins occur, and lysine is damaged by formation of unavailable lysine-sugar complexes (lactulose-lysine). The nutritional, technological and analytical aspects are discussed in detail.

INTRODUCTION

The dehydration of food is undoubtedly one of the most important achievements in man's history, making him less dependent upon a daily food supply even under adverse environmental conditions. Food processing and particularly drying are essential to our civilisation, and they provide many advantages in respect of food hygiene, distribution, storage and convenience. Mild heat treatment often improves digestibility by denaturating **unavailable** proteins and carbohydrates, and by inactivating certain proteinase inhibitors or vitamin coupling proteins. Damage can be caused however, if higher temperatures or longer treatment times are used.

Substantial nutritive losses are caused to vitamins by heating, and results for thiamine are summarised in table 1 (ref. 1).

TABLE 1

Losses in thiamine (Vit. B_1) during drying(ref. 1)

Product	Conditions	Losses in %
Pork	freeze dried, -40 C	5
Pork	air dried	50-70
Chicken	freeze dried, 1000 μHg	5
Beef	freeze dried	5
Beans[a]	air dried	5
Cabbage[a]	unknown	9
Corn[a]	unknown	4
Peas[a]	unknown	3

[a] Does not include blanching losses

Several amino acids are also damaged by drying. The influence of processing on protein quality which has been well known for many years, was first reported by McCollum and Davis in 1915 (ref. 2) for heated milk. The effect can be measured, and is expressed as a decrease in digestibility and biological value of the protein, (table 2), (ref 3).

Lipid quality is also affected by dehydration, and this aspect is discussed in another paper in this symposium. In addition, the utilization of minerals can also be impaired by drying as recently reported (ref. 4). However, carbohydrates are not adversely affected to any great extent, and in fact, the gelatinisation of the starch may improve its digestibility and energy utilisation.

TABLE 2
True digestibility (TD), biological value (BV) and net protein utilisation (NPU) of casein and a heat treated casein for rats (ref. 3)

	TD	BV	NPU
Casein	95 ± 1	85 ± 4	81 ± 4
Casein + 20% glucose, heated (drying at 105°C, 24 h)	75 ± 5[a]	62 ± 6[a]	46 ± 7[a]

[a] Six rats per group, all differences were significant ($p < 0.05$)

Nutritional damage to food is of practical significance only if the complete daily diet does not provide an adequate intake of the nutrient in question. In regard to vitamins, there are few major problems except for thiamine (Vit. B_1), folic acid or pyridoxine (Vit. B_6) (ref. 5). The addition of sulfur dioxide to preserve color in some plant foods increases thiamine loss. Otherwise the loss of most other B-vitamins seems to be of the same magnitude as thiamine (table 1). In general, loss of B-vitamins are usually less than 10%, and dried foods do not greatly contribute to our supply of thiamine, folic acid and vitamin B_6 (ref. 5). Vitamin C is affected to a higher degree, but is not limiting in the average diet (ref. 5).

However, amino acids which are partially inactivated or destroyed are less easily substituted, especially if a particular amino acid is limiting in the diet. This is especially true of lysine, which is sensitive to heat damage and is in short supply in cereal proteins, a fact which can be important in developing countries where high quality animal proteins may be scarce and expensive. Further difficulties can be caused if food protein reactions create amino-acid imbalance or formation of lysinoalanine and similar compounds.

Many reactions lead to derivatives of amino acids which can not be detected by normal chemical analysis. The reactions mainly responsible for protein damage were discovered in the last twenty years, but a clearer understanding of the influence of heat on the availability of amino acids is still required. For this reason the main focus of this presentation will be on protein quality.

EXPERIMENTAL

Studies were conducted on model proteins, maize drying and milk drying.

Crude protein, amino acids including furosine and lysinoalanine. Total nitrogen was determined by macro Kjeldahl digestion, and protein estimated as N x 6.25. Amino acid values were obtained from conventional amino acid analyses carried out on 200-500 mg samples, following either digestion with 600 ml of 6M HCL under reflux, or 50 ml 6M HCl in Pyrex glass botles at 110° C for 24 hours. A Biotronic (München, F.R.G.) LC 6000 and a Kontron (München, F.R.G.) Liquimat III were used for analysis. Cystine/cysteine and methionine were determined in the same way but using an oxidising step with 100 ml performic acid (5 ml hydrogen peroxide, 30% + 95 ml formic acid 98% + 100 mg phenol) for 16 hours at 0°C before hydrolysis. Cystine/cysteine were then determined as cysteic acid, and methionine as methionine sulphone. Furosine was determined in the same way as the amino acids (without oxidation) but 300 ml 7.8 M HCl were used instead of 6M HCl. Furosine is a reliable indicator of fructoselysine (or lactuloselysine or maltuloselysine) the most important intermediate of the initial stage of the Maillard reaction (refs. 6, 7, 8). Details of the analytical procedure and calculation of inactivated and available lysine have been described elswhere (refs. 9, 10). Lysinoalanine (LAL) was determined in the same way as the amino acids with the exception that larger samples were used to obtain measurable concentrations. The mechanism of the formation of LAL is shown in figure 1.

Fluorodinitrobenzene (FDNB)-reactive lysine: The direct FDNB method of Carpenter (ref. 11) was used with some further modifications (refs. 12, 13)

Portal plasma test: Adult rats adapted to a high protein (80%) diet and trained to consume their meals within 15 minutes were used. The concentrations of the free amino acids in the plasma of the portal vein were measured 45 and 60 minutes, or 45, 60, 75, 120 and 180 minutes after feeding a test meal of 2 to 3 g. Relative amino acid availability was calculated from the difference of concentrations between the fasting levels and the levels at different times after the uptake of the test meal. The experimental method has been described in full (ref. 14, 15).

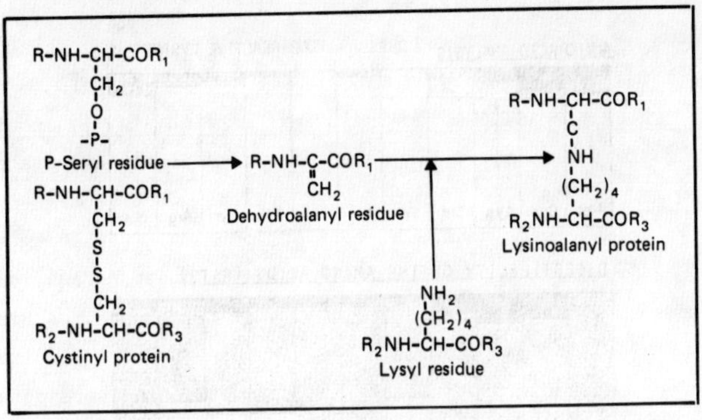

Fig. 1: Formation of lysinoalanine from P-serine or cystine with lysine

Digestibility trials and feeding experiments with chicken: Some samples were analyzed for their digestibility in rats, and the biological value was also obtained in one case (see table 1). The methods used are mentioned in the literature (ref. 16) and are described in more detail elsewhere (refs. 14, 15, 36). The feeding experiments were conducted according to normal practice (refs. 18, 19).

MODEL EXPERIMENTS ON THE MECHANISMS OF NUTRITIONAL DAMAGE

Amino acids measured in amino acid analysis are only partially utilisable by the organism. The discrepancy between total amino acids and the amount actually utilisable is difficult to calculate and leads to errors in predicting the protein quality of a food. In general, it can be accepted that in a mildly to moderately heat-treated protein, only a small amount of amino acid is destroyed and can not be determined by chemical analysis. The inactivation of amino acids can however, be substantial. In severely damaged proteins on the other hand, amino acid destruction is more significant. This is demonstrated using the example of an isolated soy protein, heated at two temperatures (fig. 2). In both cases, the amino acid analysis exhibited little damage, and only the FDNB-reactive lysine showed appreciable changes. But the biological tests however, made it clear that considerable losses in nutritive value had occurred. Although amino acid destruction was less than 20%, the

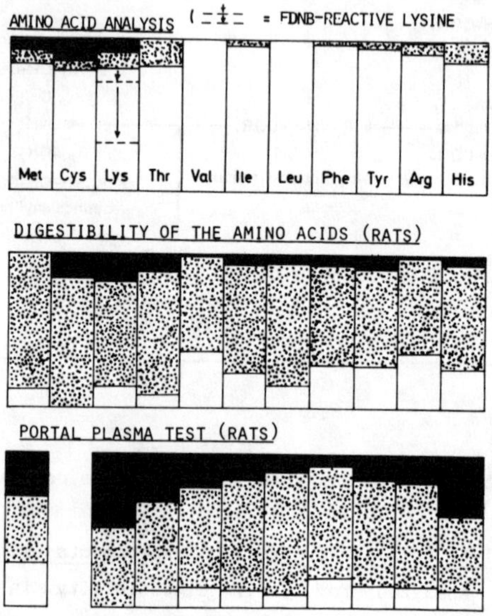

Fig. 2: Losses of essential amino acids as measured by different methods in an isolated soya protein. (Relative values; unheated sample = 100%). Black = losses at 138°C; Dotted = losses at 160°C (both drying for 24 hours).

actual protein quality was nearly zero in the more damaged sample.

Since most of the amino acids in proteins are protected by the peptide bond, only free amino acids and certain amino acids with free functional groups can be affected, unless the complete protein molecule is destroyed. Apart from total destruction, the main reactions are derivatisation of amino acids and their crosslinking with others within the protein molecule. Oxidative reactions, namely of the sulphur-containing amino acids, and reactions of free amino groups with reducing sugars in the Maillard reaction lead to more or less unavailable derivatives, but not to crosslinking. Reactions with aldehydes, acid anhydrides or phenolic compounds lead to derivatives of different availability and to crosslinking, as demonstrated by the reaction of lysine with caffeic acid (ref. 20). Crosslinking occurs between lysine on the one hand and glutamine or asparagine on the other,

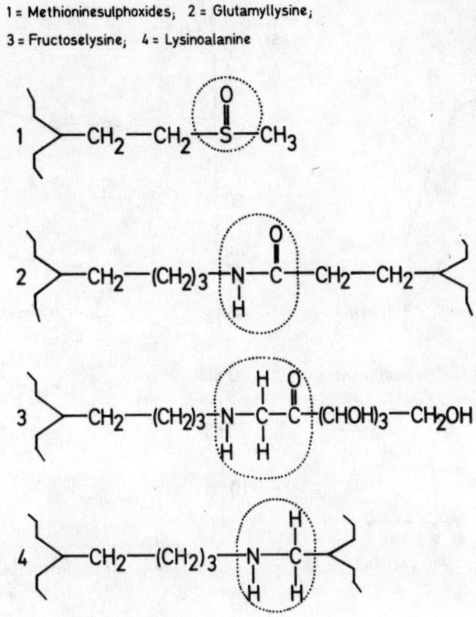

Fig. 3: Lysine reactions during food dehydration.

leading to products which are still available. Lysinoalanine is a product of cross-linking which is almost unavailable (refs. 21, 22). Figure 3 shows some reactions which can be expected during food drying.

While reactions usually influence the digestibility of protein to some extent, crosslinking, if it occurs extensively, always impairs digestibility (fig. 2), even though only a few amino acids are involved and could become available if liberated during digestion.

The principal reason for reduction in protein quality is the Maillard reaction between free amino groups in the protein and reducing sugars, which leads to unavailable amino acid-sugar complexes such as the fructoselysine moiety and may cause the destruction of both the sugars and the amino acids. Since the Maillard reaction is the subject of another paper in the symposium it will not be discussed here, although some relevant facts will be presented in relation to milk drying.

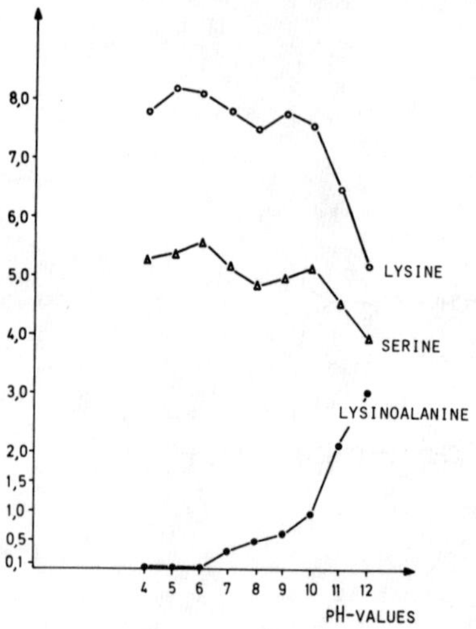

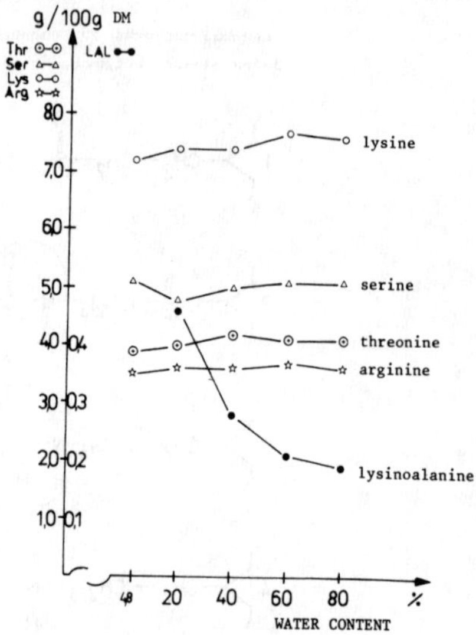

Fig. 4: Effects of pH conditions on the formation of LAL during air-drying of casein (24 h 90° C). Each point:- 4 values.

Fig. 5: Influence of dry matter content on formation of LAL in heating casein (90 , 24h, pH 7.0); difference 20% vs. 60 + 80% water is significant.

Racemisation of certain amino acids due to heat or alkaline treatment, can also occur. The amino acids mainly involved are aspartic acid, phenylalanine, glutamic acid, alanine and leucine (refs. 23, 24). However, it is unlikely that this phenomenon occurs to any appreciable extent during food dehydration.

A compound which can be found in considerable amounts in some dried foods is lysinoalanine (LAL) (refs. 25, 26). Although usually formed under alkaline conditions, it can also be found in a neutral environment, namely milk products. The fact that lysinoalanine causes cytomegaly in the epithelial cells of certain regions in the kidneys of rats has given rise to discussion about its toxicity (ref. 27). Its formation depends on pH conditions (fig. 4). As in all cases of heat damage, higher temperatures accelerate the reaction, although there is a significant decrease at temperatures above 100°C. Figure 5 shows the

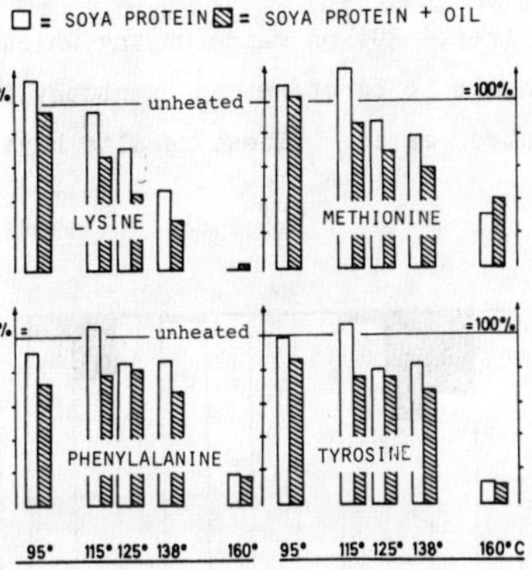

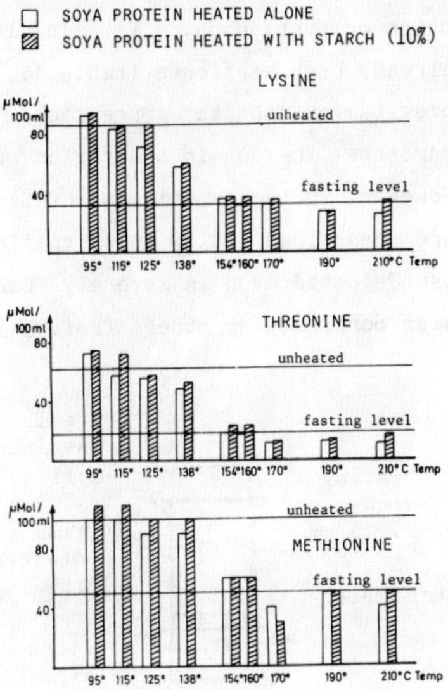

Fig. 6. Relative concentrations of several amino acids in the portal plasma of rats 45-60 min after a test meal of heated proteins(unheated=100; without and with oil; ref. 28)

Fig. 7. Concentrations of several amino acids in the portal plasma of rats 45-60 min after a test meal of heated proteins(without and with starch; ref. 29)

the effect of water content on LAL formation. This is significant because during drying the product passes through various stages of temperature and moisture content.

As mentioned earlier crosslinking reactions impair the digestibility of protein and the availability of amino acids (fig. 2). The effect of certain food constituents on the degree of damage is also of interest. Figures 6 and 7 show results of experiments in which concentrations of free amino acids in the portal plasma of rats were measured after test meals of proteins heated together with fat or starch (refs. 28, 29). While fat aggravated the heat damage, starch had some protective effect. This was noteworthy because the starch decomposed to reducing sugars during the course of heat treatment, especially in the region of 170 - 200°C.

MAIZE DRYING

Plant materials are often dried in single-stage driers, either batch or continuous. Vitamin losses in plant food during drying have already been mentioned (table 1), and losses of fat soluble vitamins and previtamins can be appreciable. Bluestein & Labuza (ref. 1) found carotene losses in the region of 10 - 40 % during drying of carrots. However, in the experiments of Strehler (ref. 30) on maize drying which are described below, no significant damage to carotene and xantophyll was detected even in severely heat-treated maize. These results have been confirmed by others (ref. 31).

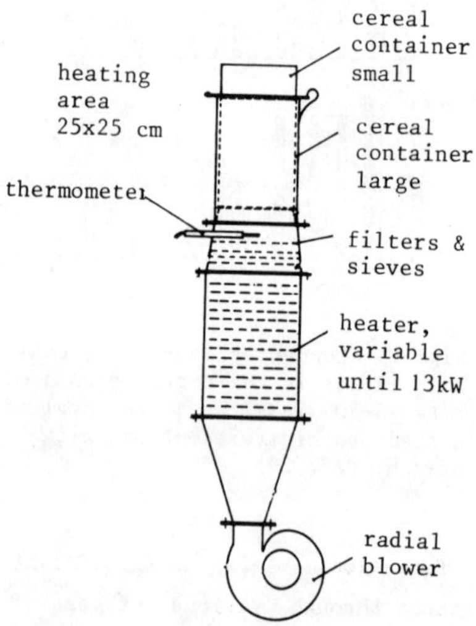

A. Strehler,1974

Fig. 8. Schematic drawing of the pilot dryer used by Strehler(ref.30)

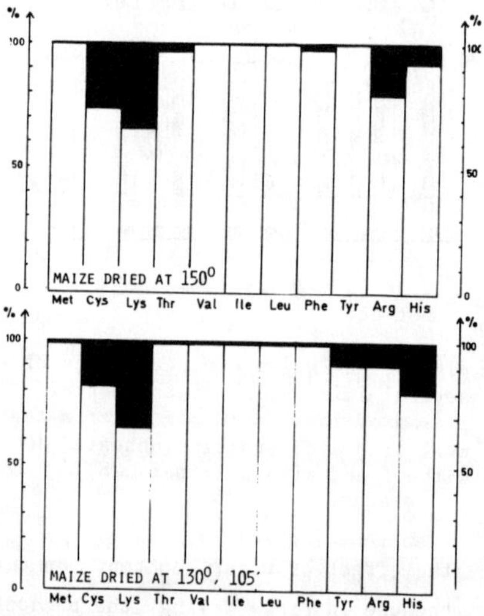

Fig. 9. Losses in essential amino acids (black) during drying of maize (carefully dried maize samples = 100) (from: ref. 31)

Figure 8 shows the pilot dryer used by Strehler in his experiments (ref. 30). Working in cooperation with Strehler our laboratory in Munich carried out the amino acid analyses and the experiments with quail chicken and broiler chicken (refs. 18, 30 and unpublished). Several amino acids were significantly damaged at high drying temperatures. Figure 9 shows results of analyses on four maize samples,

TABLE 3. Drying conditions, lysine content and weight gains of broiler and quail chicken on several maize samples(classified on increasing lysine)

heating conditions				lysine	weight gains of the chicken			
temp. °C	time,min	air (m/s)	layer[a]	(g/16gN)	broiler(rel) g/10days		quails(rel) g/3 weeks	
135°	85	0.45	0- 5	2.4	59±7	(69)	3±5	(0)
140°	55	0.30	0-15	2.8	70±5	(81)	9±6	(55)
--[b]	--	--	--	2.8	70±6	(81)	--	--
100°	60	0.48	0-16	2.9	73±7	(85)	--	--
100°	120	0.26	0-16	3.0	70±6	(81)	15±3	(88)
140°	25	0.30	0- 7	3.1	75±6	(87)	--	--
85°	180	0.50	--	3.1	84±8	(98)	--	--
60°	120	0.70	0-15	3.2	79±4	(92)	--	--
90°	240	0.30	0- 5	3.2	80±8	(93)	--	--
135°	85	0.45	20-25	3.3	84±7	(98)	--	--
60°	120	0.70	0-15	3.4	75±5	(87)	--	--
140°	13	0.30	0- 4	3.4	79±6	(92)	17±4	(100)
60°	120	0.70	0-15	3.4	81±5	(94)	--	--
60°	390	0.30	0-46	3.6	84±8	(98)	16±3	(95)
90°	240	0.30	20-30	3.6	86±8	(100)	--	--
70°	270	0.30	0-45	3.7	84±10	(98)	--	--

[a] zone in the dryer, measured from the heating inlet
[b] not determined

of which two were heat damaged. Losses of cystine and the basic amino acids were noted. Lysine loss was appreciable. A total of 118 maize samples were dried under different conditions, of which 16 were tested in animal experiments. Table 3 gives (a) heating conditions, (b) lysine content and (c) results of the feeding tests with chicken. There was a highly significant correlation between lysine content and weight gain of the male broiler chicken (r = 0.88). The main factors causing damage were temperature, time of heating, air velocity and the thickness of the maize layer in the drier.

Another 42 maize samples, dried in the lowest zone in the dryer (0-5 cm) were selected to determine the correlation of lysine content with heat intensity. Temperature and heating time were combined to give "heat intensity", based on heating time (minutes) at 100°C. A multiplication factor of 1.5 per 10°C increase or decrease of temperature was used, e.g. 10 minutes at 110°C corresponded to 15 minutes at 100°. As can be seen from figure 10, a good correlation was achieved. The effect of heat intensity on lysine contents and weight gains of broiler chicks is given in figure 11. Correlation is quite good.

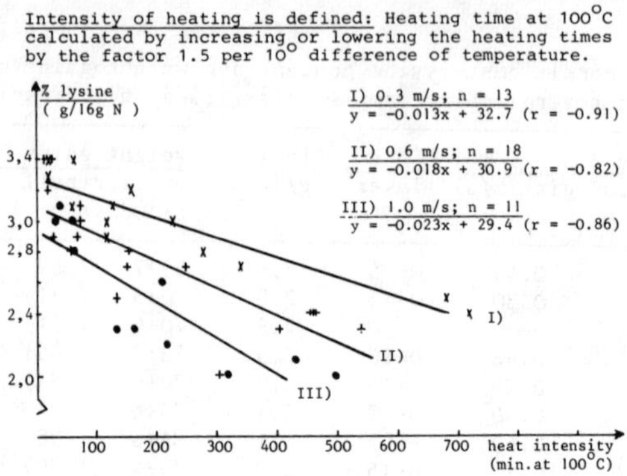

Fig. 10: Effect of heating intensity on lysine content in maize, at different air velocities.

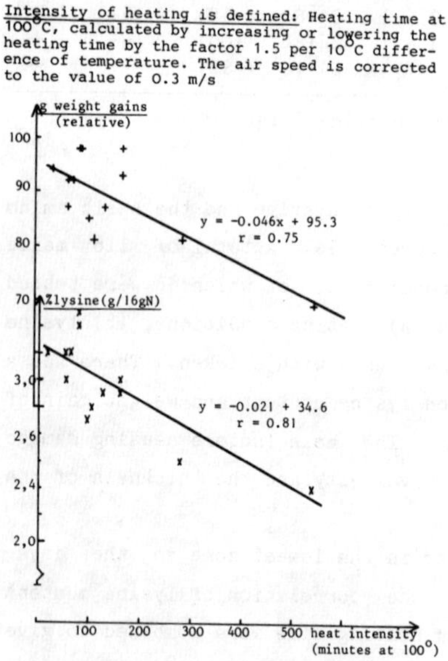

Fig. 11: Effect of heating intensity on the weight gains of broiler chicks and on lysine content in maize.

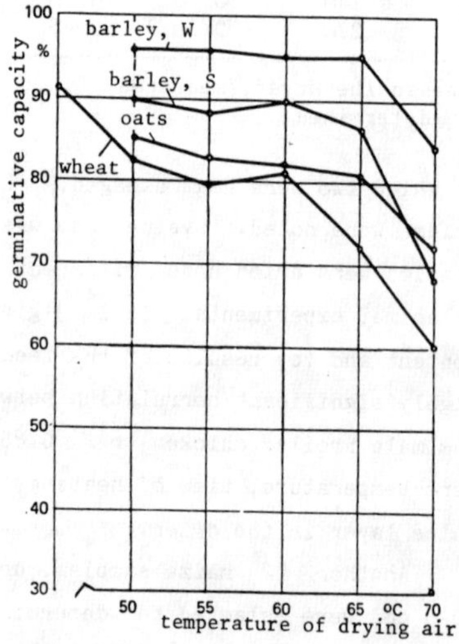

Fig. 12: Influence of drying temperatures on the germinative capacity of several cereals (ref. 32).

An aspect of nutrition often mentioned by health food enthusiasts is the germination capacity of grains. Results of experiments on wheat, oats and barley are given in figure 12 (ref. 32).

Regarding the Maillard reaction and the mechanisms of heat damage discussed earlier, table 4 gives results of furosine and lysinoalanine determinations in several dried foods. In the maize samples discussed earlier only small amounts of furosine were detected, viz. in the maize that was not fully ripened.

TABLE 4.

Contents in lysine, inactivated lysine (given in g/16g N) and lysinoalanine (LAL in ppm = mg/160g N) in several foods. (from refs. 3, 25, 27, 33)

Products	n	lysine	inactivated[a] lysine	(% losses)[b]	LAL ppm
crisp bread	1	3,5	0.3	7	0
zwieback	1	0.9	1.4	61	70
biscuits	1	4.2	0.4	10	0
butter cookies	1	1.5	0.3	19	0
ginger bread	1	1.9	2.1	53	0
pretzels	1	2.5	0.6	19	47
noodles and other items	7	4.0-5.4	0.1-1.8	3-26	0-138
potato chips	1	7.6	1.4	16	0
corn flakes	1	1.1	0.1	8	750
skim milk, spray dried	37	8.5	0.6	6	tr[c]
skim milk, roller dried	42	8.1	0.9	10	tr
whey, spray dried	33	8.3	0.6	7	nd[c]
whey, roller dried	16	5.9	3.9	40	nd
baby food, dried	44	8.1	0.7	8	tr
baby food, liquid, sterilized	23	7.6	0.7	8	270-920
chocolate	12	5.7	1.5	21	tr

[a] calculated from furosine and (cereals) pyridosine
[b] lysine losses = inactivated lysine in % of determined lysine + inact.lys.
[c] tr = traces (50 ppm); nd = not determined, but high amounts not possible

DRYING OF MILK

There are many steps involved in the dehydration of liquid foods such as milk. For milk these steps are: preheating and sterilisation, preconcentration, drying and storage. Since the mechanism of heat damage caused by preheating and storage is very complex, (refs. 3, 8) they will not be discussed here because of space limitation.

Vitamin losses in milk during drying are similar in magnitude to those of other foods, (table 5).

TABLE 5

Losses of some vitamins during milk drying (ref. 35)

	Spray drying	Roller drying
Thiamine (Vit. B_1) losses in %	10	15 - 20
Cobalamine (Vit. B_{12}) losses in %	20 - 30	30 - 40
Ascorbic acid (Vit. C) losses in %	20	30 - 60

Since milk proteins are relatively rich in lysine, and lactose is a reducing sugar, lysine is expected to be damaged. Figure 13 shows results of amino acid analyses in two roller dried skim milks, one heated gently and one overheated. The overheating was carried out by increasing the steam pressure of the drum from 1.9-2.1 bar to 3.2 bar and reducing the speed of rotation from 26-28 to 24 revolutions per minute. The result was losses in lysine of 30%, cystine 18% and histidine 12%. All other amino acids were only slightly affected.

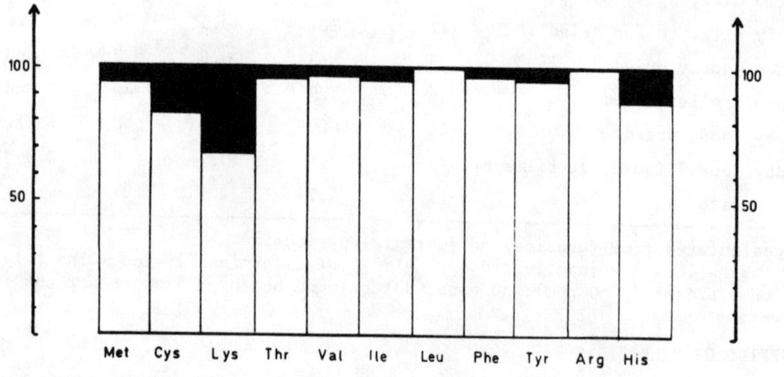

Fig. 13: Losses in essential amino acids during drying of a skim milk sample under severe conditions (see text) compared to a sample carefully processed from the same raw milk.

The FDNB reactive lysine showed similar damage. From the furosine values, inactivated lysine was calculated to be 3.2% which indicates a loss in available lysine of nearly 40%.

While severe lysine damage is caused by the reaction of lysine to lactuloselysine (galactose-fructoselysine) and is also caused in the advanced stages of the Maillard reaction, other amino acids with the exception of cystine are only scarcely damaged. This was demonstrated by an experiment in which lysine-HCl was added as a supplement to a diet containing severely heat-damaged dried skim milk which contained only 2.2% available lysine, representing a loss in available lysine of 75%. The supplementation of lysine restored the protein quality almost to the level of the control sample of good quality. The results of the experiment are given in figure 14 (ref. 36).

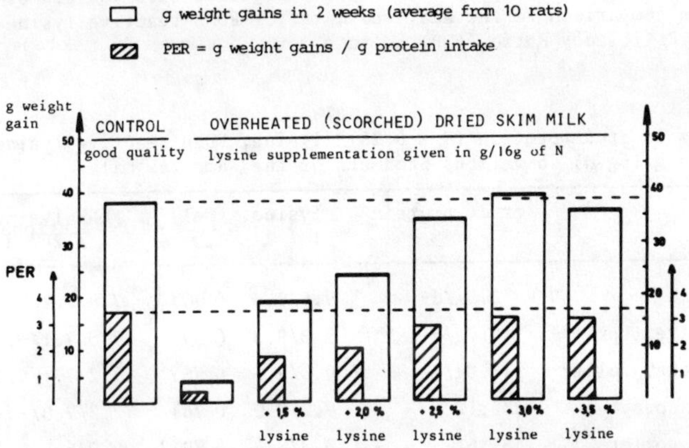

Fig. 14. Effects of lysine supplementation of a severely heat damaged dried skim milk on weight gains and Protein Efficiency Ratio(PER) of rats.

Similar results were obtained in experiments with casein which was dried together with 10% glucose at 90°C for 24 hours (initial water content 50%). The portal plasma test indicated significant damage to lysine only, and digestibility studies supported this result. Furosine levels indicated a 40% loss of available lysine, and 45% loss of FDNB-reactive lysine. Figure 15 shows the results of experiments on this heat-damaged casein (ref. 14).

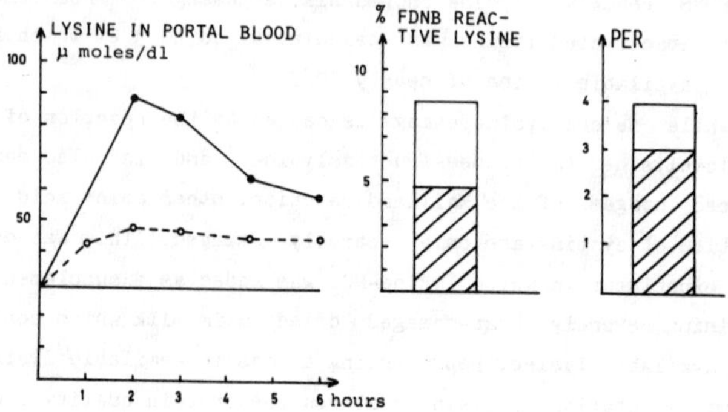

Fig. 15. Influx of lysine into the portal plasma of rats after a test meal of casein (——) or heated (dried at 90°C together with 10% glucose) casein (---) in comparison to the analysed data for FDNB reactive lysine and the Protein Efficiency Ratio (PER) in rats.

TABLE 6

Contents in crude protein (N x 6.25), lysine, FDNB reactive lysine(both given in g/16g N) in various products of the same raw milk

	crude protein	lysine 1 / 2	(rel)	FDNB-lysine 1 / 2 / 3	(rel
skim milk	3.5g/dl	9.6/9.3	(100)	7.6/7.2/-.-	(10
condensate, spray	14.7g/dl	9.3/9.3	(99)	7.5/7.1/-.-	(9
condensate, roller	16.8g/dl	9.0/9.2	(96)	7.2/7.1/6.9[a]	(9
powder, spray	35.1%	9.2/9.0	(96)	7.2/7.0/-.-	(9
powder, roller	35.6%	8.3/8.4	(89)	6.9/6.5/6.4[a]	(9

[a] These values were not used for the average, which was the basis for the relative values

Table 6 shows the influence of preconcentration and drying of mil on the content of FDNB reactive lysine. The results show that bot processes affect lysine concentration. The three samples were obtaine from different plants and analysed at least in duplicate.

Freeze drying is assumed to be one of the most gentle of dryin methods. To reduce drying time, heating platen temperatures ar sometimes adjusted to higher levels than recommended. In th dissertation of Busch-Stockfisch (ref. 37), experiments on freeze drie skim milk powders were conducted, of which some results are given i

TABLE 7.

Influence of freeze drying (17-24 hours at 0.5 torr) on amino acid content and Protein Efficiency Ratio (PER) of rats with different diets (ref. 37)

	temperature of freeze drying in °C		
	40	55	70
Lysine[a]	8.6	8.5	8.3
Furosine[a]	0.05	0.11	0.22
Available lysine[a]	8.6	8.4	8.1
Methionine + cystine	3.6	3.6	3.1
PER, milk alone	3.51 ± 0.21	3.56 ± 0.14	3.46 ± 0.25
PER, diet low in lysine[b]	2.85 ± 0.12*	nd[d]	2.71 ± 0.13*
PER, diet low in met + cys[c]	3.57 ± 0.14**	nd	3.25 ± 0.16**

[a] the determinations were made in our laboratory in Munich
[b] 60% of protein based on wheat, maize, sesame meal and 40% from the dried milk. The basal diet alone yielded a PER of 1.30
[c] 50% of protein based on linseed and 50% from the dried milk. The basal diet yielded a PER of 3.32
[d] nd = not determined
* = significant at the 0.05 level
** = significant at the 0.01 level

in table 7. The lysine and furosine determinations of this work were conducted in the author's laboratory. As can be seen from table 7 temperature had an effect on the concentrations of lysine, methionine and cystine, and also had a significant effect on the growth of rats fed with milk in diets deficient in lysine, methionine and cystine.

To conclude this section on milk drying, it can be stated that most of the important heat damage mechanisms have now been clarified i.e. in the Maillard reaction, (see relevant paper in this symposium and also refs. 3, 8, 9, 11, 34), and it is now possible to produce recommendations for product manufacturing to minimize product damage.

CONCLUSIONS

As with many other preservation processes, drying can decrease the nutritive value of foods. In this paper only the effects on vitamins and protein quality were considered. Vitamin losses are significant in those compounds which are known to be limiting in our diet. On the other hand vitamin losses are often lower than the natural fluctuations in plant foods which depend on variety, location, fertilization, etc. and on harvesting and cooking methods (ref. 38).

In contrast to vitamins, the supply of proteins cannot be increased at will, and animal proteins rich in lysine are often in short supply.

By using modern sensitive and specific analytical methods, very small changes in foods can be detected, as in the case of ultra high temperature (UHT)- treated milk (refs. 8, 34). While many of these changes have little nutritional significance, they can be used as an indicator of good manufacturing practice and product quality. Quality control should be used not only as a tool for supervising production, but also for improving processing equipment and procedures and for evaluating new concepts. The improvement in the quality of roller dried skim milk which the author has observed for a quarter of a century, bears testimony to this fact.

REFERENCES

1 P.M. Bluestein and T.P. Labuza, in R.S. Harris and E. Karmas (Ed.), Nutritional evaluation of food processing, The Avi Publishing Comp., Westport, 1975, pp 289-323.

2 E.V. McCollum and M.J. Davis, J. Biol. Chem. 23(1915)247.

3 H.F. Erbersdobler, Kieler Milchwiss. Forschungsber. 35(1983)301-311.

4 P.A. Finot, E. Magneat and D.E. Furniss, in Proceedings of the Third Int. Symp. on the Maillard Reaction, (1985) in press.

5 Ernahrungsbericht 1984, Deutsche Gesellschaft für Ernährung, Frankfurt (Ed. and Publisher), 1984.

6 J. Brüggemann and H. Erbersdobler, Z. Lebensm. Unters. Forsch. 137(1968)137-143.

7 H.F. Erbersdobler, in M. Friedman (Ed.), Protein Crosslinking-B, Nutritional and Medical Consequences, Plenum Publ. Corp., New York, 1977, pp 367-378.

8 H.F. Erbersdobler, in Proceedings of the Third Int. Symp. on the Maillard, (1985), in press.

9 H.F. Erbersdobler, B. Holstein and E. Lainer, Z. Lebensm. Unters. Forsch. 168(1979)6-8.

10 A. Brandt and H. Erbersdobler, Landw. Forschung 28./II. Sonderheft, (1972)115-119.

11 K.J. Carpenter, Biochem. J. 77(1960)604-610.

12 H. Erbersdobler and H. Zucker, Z. Tierphys. Tierern. Futtermittelk. 19(1964)224-255.

13 V.H. Booth, J. Sci. Food Agric. 22(1971)658-664.

14 H. Erbersdobler, Thesis (Habilitationsschrift), Univ. of Munich, 1967.

15 H. Erbersdobler, H. Dümmer and H. Zucker, Z. Tierphys. Tierern. Futtermittelk. 24(1968)136-152.

16 United Nations University, P.L. Pellett and V.R. Young (Ed.), Nutritional Evaluation of Protein Foods, The United Nations University, Tokyo (1980).

17 H. Erbersdobler, J. Gropp and H. Zucker, Wiss. Z. Univ. Rostock 18M1/2(1969)145-150.
18 H. Erbersdobler and J. Gropp, Archiv f. Geflügelkunde, 35(1971)36-40.
19 H. Erbersdobler, J. Gropp and H. Beck, Proc. Nutr. Soc. 34(1975) 21-28.
20 R.F. Hurrell, P.A. Finot and J.L. Cuq, Br. J. Nutr. 47(1982)191-212.
21 J. Bjarnason and K.J. Carpenter, Br. J. Nutr. 23(1969)859-868.
22 B.J. Struthers, J.E. Brielmaier, M.L. Raymond, R.R. Dahlgreen and D.T. Hopkins, J. Nutr. 110(1980)2065-2077.
23 F. Hayase, H. Kato and M. Fujimaki, J. Agr. Fd. Chem 23(1975)491.
24 M. Friedman, J.C. Zahnley, P.M. Masters, J. Food Sci. 46(1981)127-131.
25 R.J. Fritsch and H. Klostermeyer, Z. Lebensm. Unters. Forsch. 172 (1981)440-445.
26 M. Sternberg and C.Y. Kim, in M. Friedman (Ed.), Protein Crosslinking-B Nutritional and Medical Consequences, Plenum Press, New York, (1977)73-84.
27 H.F. Erbersdobler and E. Trautwein, Ernährungs-Umschau, 31(1984)147-150.
28 G. Weber, med. vet. Dissertation, Univ. of Munich (1972).
29 F. Grimm, med. vet. Dissertation, Univ. of Munich (1973).
30 A. Strehler-Pöppel, agr. Dissertation, Technical Univ. of Munich (1974).
31 A.H. Jensen, S.W. Terrill and D.E. Becker, J. Animal Sci. 19(1960) 629-638.
32 A. Strehler, R. Schäfer, H. Fuchs and H. Scheller, Grundl. Landtechnik, 28(1978)108-112.
33 J. Steinig and A. Montag, Z. Lebensm. Unters. Forsch. 174(1982) 453-457.
34 H.F. Erbersdobler, H. Reuter and E. Trautwein, in Royal Soc. of Chem. (Ed. & Publ.) Challenges to Contemporary Dairy Analytical Techniques (1984)328.
35 E. Renner, Milch und Milchprodukte in der Ernährung des Menschen, Verlag Th. Mann KG, Gelsenkirchen, VV-GmbH, München (1982).
36 H. Erbersdobler, and H. Dümmer, Z. Tierphys. Tierern. Futtermittelk., 28(1971)224-231.
37 M. Busch-Stockfisch, Troph. Dissertation, Univ. of Bonn, (1979).
38 H.F. Erbersdobler, H.-G. Greulich and A. Gross, Schriftenreihe der Agr. Wiss. Fak. d. Univ. Kiel, Paul Parey, Hamburg and Berlin, 63(1982)201-207.
39 E. Trautwein, S. Trapp and H.F. Erbersdobler, Ernährungs-Umschau, 32(1985)227.

MICROBIOLOGICAL QUALITY OF DRIED FOODS

P.A. GIBBS
Leatherhead Food R.A., Leatherhead, Surrey (U.K.)

ABSTRACT

The basic condition necessary for the enzymic or microbiological spoilage of foods is the presence of water. Limitation of water, either by adding solutes or by physical removal of water, minimises microbial growth and enzymic activity. However, drying is also an effective means of preserving micro-organisms in a viable state, even though their numbers may be reduced and a proportion sub-lethally damaged. Survival during and after drying will depend upon the physico-chemical conditions experienced by the micro-organisms, viz. temperature, A_w, pH, preservatives, oxygen, etc. The survival of food-spoilage organisms may give rise to problems in a reconstituted food item, but survival of food-borne pathogens must be viewed much more seriously.

In order to minimise organoleptic changes in foods during drying, times and temperatures are kept as short and as low as feasible. Thus microbes are not exposed to highly lethal conditions. To ensure the safety of dried foods, some form of lethal treatment, usually heat, is often applied prior to drying. With certain foods, e.g. dried pasta or vegetables, this is not feasible and pre-processing or selection of ingredients to minimise the occurrence of pathogens must be undertaken, together with hygienic handling and as rapid drying as possible.

The conditions of storing and handling of post-pasteurised foods must not be such that microbes can contaminate, grow or form toxins in the food prior to drying, for example in surge tanks or pre-concentrators. The role of microbial enzymes, particularly those that are heat-resistant, in causing spoilage must also not be forgotten. Good Manufacturing Practice (GMP) guidelines must be established for each type of operation, together with recommended microbiological protocols.

The microbiologist's problems in sampling and analysing dried foods are exacerbated by the non-uniform distribution of and by sub-lethal damage in the surviving microflora. Obtaining representative samples for microbiological analysis is statistically a difficult problem, especially since drying of foods is usually on a very large scale. Comparatively little reliance can be placed on finished product microbiological analysis, except in detecting quite gross survival or contamination. If sub-lethal damage in the surviving microflora from heat or low A_w, is not repaired prior to the use of highly selective media for the detection of food-poisoning microbes, a false sense of security in the process and product may be engendered. The problem of appropriate analyses is also frequently discussed; in general a "total viable count" of the type of organisms important in that food product is carried out, together with analyses for indicator organisms (e.g. "coliforms") and for specific pathogens, e.g. Staph.aureus, Salmonella, Clostridium perfringens. Certain of these analyses should be related to the likely end use of the dried food.

INTRODUCTION

The oldest method of preserving foods against microbiological spoilage is by drying. The earliest recorded farming communities, of ancient Egypt, preserved grains by drying and storing them in granaries, protecting them against rehydration. Drying of meat and fish by smoking and salting, and sun-drying of fruits and vegetables undoubtedly occurred long before recorded history. Dried plants, as hay, are also traditional winter fodder for cattle in those countries where long spells of warm dry weather can be relied upon. The basis for the stability of dried foods over long periods is that the presence of water is essential for enzymic reactions; removal or limitation of water prevents endogenous enzymic degradation of foods and also the activities of contaminating micro-organisms. Removal of water from foods results in an increase in solute concentration in the remaining moisture, limiting the water available for enzymic reaction. Solute concentration can also be increased by addition of salt or sugar to a food. The level of available water or water activity (A_w) has a profound effect on the growth of micro-organisms (Table 1); the growth of a restricted range of

TABLE 1

Water activity and growth of micro-organisms in foods

A_w	Micro-organisms inhibited	Examples of foods
0.95	Pseudomonas, Escherichia, Shigella, Bacillus, Cl.perfringens	Fresh milk, vegetables, fish, meat, fruits
0.91	Salmonella, Cl.botulinum, Lactobacillus, Pediococcus, some moulds and yeasts (Pichia, Rhodotorula)	Some cheeses (Cheddar, Swiss, Muenster), cured ham, fruit concentrates (55% m/m sucrose or 12% NaCl)
0.87	Candida, Torulopsis, Hansenula, Micrococcus	Dry cheeses, fermented meats (salami), sponge cakes (65% m/m sucrose or 15% NaCl)
0.80	Most moulds (incl. Penicillium) Staph.aureus, Zygosacch.baillii (most) Debaryomyces	Fruit concentrates, syrups; sweetened condensed milk, flour, rice etc. Fondants, fruit cakes
0.75	Aspergillus spp., halophilic bacteria	Jams, marmalade, glacé fruits, marzipan
0.65	Xerophilic moulds (Asp. chevalieri, A.candidus, Wallemia sebi) Sacch. bisporus	Fudge, nougats, jelly, molasses and raw cane sugar, dried fruits, nuts
0.60	Osmophilic yeasts (Sacch. rouxii), and moulds (Asp. echinulatus, Monascus bisporus)	Honey, toffees, caramels, dried fruits
0.50	All	Pasta of *ca* 12% moisture, spices of *ca* 10% moisture
0.40	All	Whole egg powder of *ca* 5% moisture
0.30	All	Biscuits, crackers etc. of *ca* 3-5% moisture
0.20	All	Whole milk powder of *ca* 2-3% moisture; dried vegetables of *ca* 5% moisture etc.

microbes is often characteristic of the A_w of the food. However, the type of solute may affect the lowest A_w at which a micro-organism can grow. For complete microbiological stability the A_w must be less than 0.60.

MICROBIAL RESISTANCE TO DRYING

Drying of Liquid Products

The process of drying, whether by freeze drying, ambient-to-warm air drying, solar drying or by high temperatures (e.g. spray- or roller-drying), is not *per se* lethal to all micro-organisms and many may survive. The survivors will to some extent reflect the initial microflora of the raw material, but, depending on the severity of the process, particularly the temperatures attained by the product during drying, the more resistant organisms are the most likely survivors, e.g. bacterial spores, yeasts, moulds and thermoduric bacteria (refs. 1-7).

However, there is considerable evidence that heat-sensitive food-poisoning bacteria, particularly *Salmonella* species, can also survive spray-drying processes. A widely distributed outbreak of salmonellosis due to *Salm.newbrunswick* in non-fat dried milk was reported in the USA (ref. 8) which gave rise to several studies on the survival of salmonellae through the spray-drying process. McDonough and Hargrove (ref. 9) found that while pasteurisation destroys salmonellae in liquid milk, higher temperatures are required for destruction in milk concentrates. It was also shown that although certain conditions of heat and moisture reduced the chances of survival during drying, total destruction could not be achieved. Nevertheless, Galesloot and Stadhouders (ref. 10) suggested that the source of salmonellae occasionally found in dried milk was incidental contamination via the large volumes of air in drying and cooling operations. However, LiCari and Potter (ref. 11) showed conclusively that while spray drying under commercial conditions killed substantial numbers of salmonellae in skim milk, in no case did it yield salmonella-free powder. The survival of salmonellae decreased as the outlet temperature of the dryer was increased, but marked differences between strains of *Salmonella* were noted; log reductions between *ca* 2 and 5 were recorded with an outlet air temperature of 121.1°C (Table 2).

TABLE 2
Effect of spray-drying temperatures on destruction of Salmonella[1]

Culture	Outlet air temperature (°C)	Log reduction of viable counts[2]
S.typhimurium	65.6	2.733
	93.3	4.383
	121.1	4.915
S.thompson	65.6	2.483
	93.3	3.992
	121.1	4.300
S.kentucky	65.6	1.549
	93.3	2.903
	121.1	4.743
S.tennessee	65.6	0.566
	93.3	1.153
	121.1	1.994

1 From LiCari and Potter (ref. 11)
2 Inoculum levels ca 1 - 3.5 x 10^7/100 g condensed milk on a dry basis, i.e. counts per 100 g of moisture-free solids.

The heating of re-wetted powder in the agglomerating process, followed by heating at 110°C for up to 5 minutes, produced even lower reductions in viable counts (ref. 11). Keogh (ref. 3), in reviewing the survival of pathogens in milk products, emphasised the need for vigilance in hygiene and in process parameters at all stages of manufacture if the occurrence of pathogens is to be minimised.

Miller, Goepfert and Amundson (ref. 12) and Espina and Packard (ref. 13) showed that the initial % total solids (TS) in spray drying of concentrated milks exerted a considerable influence on the survival of vegetative bacteria including salmonellae. An increase in TS increased the resistance of salmonellae and Escherichia coli (ref. 12) but decreased the resistance of Lactobacillus acidophilus (ref. 13). At either of the two initial TS (40% and 20%) of milk, drying to a lower moisture content (using higher inlet and outlet air temperatures) achieved a greater reduction in numbers of surviving salmonellae or E.coli (ref. 12). Chopin, Mocquot and Le Graet (ref. 14) also observed that a higher air outlet temperature, resulting in a lower final moisture content, increased the lethality of the process for Staph.aureus, E.coli and Microbacterium lacticum. In a recent series of papers Daemen (ref. 15), Daemen and van der Stege (ref. 16), Daemen, Kruk and van der Stege (ref. 17), studying the effects of TS contents of milk and whey on the thermoresistance of two bacteria (Serratia marcescens and Staphylococcus C131) and three enzymes, examined the effects of drying conditions, e.g. air temperature and TS content of the concentrate, and analysed the drying process according to stages in which the destruction occurs. An increase in the TS content generally corresponded with an increase in the thermal resistance, in both D and Z values (ref. 15). This confirmed, in a real food situation, results on thermal resistance of vegetative cells in high solute concentrations with regard to intermediate moisture foods (see refs. 18, 19 for reviews on the heat resistance of salmonellae and Staph.aureus, respectively, in high-solute concentrations). Whilst not all solutes and concentrations, however, confer increased thermal resistance to the same degree, solutes which do not readily permeate the cell membrane and thereby cause withdrawal of water from the cell increase thermal resistance most markedly. Such would appear to be the case in concentrated milk and whey, in which the major solutes are lactose and large molecular weight proteins. During spray drying of milk and whey the outlet air temperature had the most profound effect on survival of Serr.marcescens and Staphylococcus C131, permitting calculation of a pseudo-Z value, while the inlet air temperature had only a minor effect (refs. 16, 17). Even at the lowest outlet air temperature used, survival of Serr.marcescens did not exceed 2% and Staph. C131 did not exceed 50%; this lethality was ascribed to a non-thermal drying effect. Increasing the TS content from 10% to 45% increased the lethality of the process from ca 70% to 30% survival for Staph. C131 but for Serr.marcescens decreased the lethality from ca 0.09% to 1%. Clearly these two organisms react differently to solute and heat stress.

Daemen et al. (ref. 17) explained the response of Serr.marcescens, salmonellae and E.coli as being that of thermolabile organisms, and the response of Staphylococcus C131 and Lactobacillus acidophilus (ref. 13) as that of "thermostable" bacteria. The difference in response was essentially a reflection of the time, and therefore temperature, at which thermal destruction took place in the spray-drying process.

Daemen et al. (ref. 17) envisaged an inactivation curve consisting of four phases during the drying process. Stage 1 consists of a very short inactivation stage dependent probably on a temporary rise in temperature immediately on mixing the liquid and hot air. If the organisms are particularly thermolabile, a notable proportion of the total inactivation occurs in this stage. In Stage 2 the temperature of the droplet remains constant and quite low at the wet-bulb temperature of the air stream and little microbial inactivation occurs. As the A_w decreases the temperature of the particle rises above the wet-bulb temperature towards that of the outlet air temperature with increasing inactivation of micro-organisms. However, as the A_w falls, the thermal resistance increases until Stage 4 appears, during which microbial destruction almost ceases. The relative thermal resistance of microbes at high solute concentrations was used to explain the influence of TS content on inactivation during drying.

Although much research has concentrated on dried milk and milk products and on salmonellae surviving the processes, other foods and other pathogens have also been subjected to study. Raw liquid egg is a well-known source of salmonellae and is pasteurised before freezing or drying. Anellis, Lubas and Rayman (ref. 20) showed that pasteurisation at an alkaline pH value was most efficacious, but even so the reduction in numbers of salmonellae achieved by most pasteurisation conditions is only 3-4 log cycles (ref. 21), which is an adequate safety margin for most situations. Since egg is much more readily denatured by heat, both pasteurisation and drying are at lower temperatures than those used for milk. To ensure freedom from salmonellae, hot storage of the dried material in sealed containers for several days can be used with no serious effects on functional qualities. The temperatures recommended range from 57.2°C to 49°C with storage from 4 days to 20 days, respectively (ref. 21). Miller et al. (ref. 12) determined the lethality of spray drying of whole egg, egg white and egg yolk for *Salmonella typhimurium* at an inlet air temperature of 225$\pm$ 2°C and an outlet air temperature of 93$\pm$ 2°C, and obtained log reductions of 4.4, 5.7 and 6.2, respectively. Similarly, in a yeast cell concentrate (25% solids) the log reduction in numbers of *S.typhimurium* was 4.9. Miller et al. (ref. 12) comment that the drying process cannot replace pasteurisation for assuring freedom from Salmonella contamination in the final product.

Drying of Semi-solid Products

In the manufacture of a pasta product, such as spaghetti or macaroni, at no stage is a pasteurising heat treatment applied: indeed slow drying is carried out at temperatures from *ca* 35°C up to *ca* 63°C, although the product only reaches these temperatures in the later stages of drying as a result of evaporation of water. Thus there is a strong possibility for microbial growth, including pathogens, before the A_w of the product falls below the critical level for each organism. Mossel and his co-workers were clearly aware of the problem of staphylococcal enterotoxins occurring in pasta before 1970 (ref. 6) but it was not until 1974 that investigations of the growth and survival of *Staph.aureus* in pasta and the persistence of enterotoxins in the dried product were reported (ref. 22). These workers showed that in a wet macaroni dough (*ca* 30% moisture, A_w greater than 0.95), *Staph.aureus* increased by *ca* 5 log cycles at 25 or 35°C in 18-24h, although enterotoxin A was not detected until the viable count reached *ca* 10^6/g. Lee *et al*. (ref. 22) were of the opinion that enterotoxin production during normal manufacture was only a remote possibility since only a short period of time elapsed during the initial drying to an A_w of *ca* 0.9, at which enterotoxin production was inhibited. Similar results and conclusions were reached by Walsh and Funke (ref. 23) in a study of spaghetti manufacture. Nevertheless, if enterotoxin is formed in pasta, the producing organism may well be undetectable after a period of storage of the dry product (refs. 22, 23) but the enterotoxin will persist for at least one year (ref. 22). An interesting difference was noted by Walsh and his co-workers in the behaviour of *Salmonella typhimurium* and *Staph.aureus* during the manufacture of pasta. Whereas *S.typhimurium* was more susceptible to extrusion pressure (or shear rate) than to temperature during extrusion (ref. 24), *Staph.aureus* was relatively unaffected by shear rate, but was susceptible to the highest extruder temperature used, 55°C (ref. 23). During storage of dried spaghetti, the number of viable *S.typhimurium* declined rapidly over the first month and then more slowly in the next three to four months but then remained at a residual number for up to seven months at ambient temperature (ref. 24).

Hsieh, Acott and Labuza (ref. 25) evolved a model to predict the death of Staph.aureus and S.anatum during the drying of macaroni product. The two factors taken into account were temperature and A_w of the product and data were accumulated for thermal death rate at various values of A_w and temperature. From these data the kill expected for exposure to a particular temperature-A_w combination for an actual process time was estimated and by summation of these values achieved during an actual drying process the total log reduction in numbers was calculated. Good agreement was found between the predicted and actual lethality of a process. In the first half hour of preliminary drying (at ca 60°C, 70% R.H.) Staph.aureus declined by only 1.5 log cycles and four more log cycles during 5 hours finishing drying (ca 55°C, 50% R.H.), whereas S.anatum declined by ca 6 log cycles in the same process. Thus if initial populations are high (i.e. greater than 10^6/g) normal pasta drying operations will not be sufficient to render the product safe. This prediction has recently been found, unfortunately, to be true. An outbreak in the U.K. in 1984 of staphylococcal food poisoning was traced to an imported dried pasta product containing egg (lasagne), in which high counts of Staph.aureus (10^4 - 10^8/g) and the presence of enterotoxin A were demonstrated (Public Health Laboratory Service, Communicable Disease Surveillance Centre, unpublished). The possibility of increasing the drying temperature and thereby decreasing the drying time seems a practical proposition but technologically has been suspect as it can result in a brittle product. However, Seiler & Bruhn (ref. 26) reported that a high-temperature, short-time drying process (95°C; 0.08h pre-drying, 1.50 h final drying), rather than cooler but longer drying, produced the most acceptable product in their tests.

Chocolate. The extreme thermoresistance of vegetative organisms at very low A_w values has also been clearly demonstrated in chocolate, in which the moisture content is ca 0.5 - 1.0% m/m (refs. 27, 28, 29, 30). Addition of 2% m/m moisture to a chocolate mix reduced the D_{71} value from 20h to 4h (ref. 30). It is quite clear that normal conching temperatures for chocolate will do little to reduce any *Salmonella* contamination. It has also, unfortunately, been found that chocolate contaminated with very low levels of *Salmonella* can be responsible for widespread outbreaks of food poisoning, *Salmonella eastbourne* in northern USA and Canada in 1973/4 (ref. 31) and *S.napoli* in the U.K. in 1982 (ref. 32).

Considerable emphasis has been laid on the thermal resistance of pathogenic organisms in view of their importance in public health, although undoubtedly other organisms behave similarly. These will affect the shelf-life or spoilage of the product, leading to economic losses to the producer which although serious in the long term, do not have the immediate effect that withdrawal of a product on the grounds of safety would have on the manufacturer.

MICROBIOLOGICAL ANALYSIS OF DRIED FOODS

Sampling

The food microbiologist is faced with two major problems when analysing dried foods. The first of these problems is that of obtaining truly representative samples of a production batch or lot such that the results of any microbiological analysis on the samples present as true a picture as possible of the microbiological status of the batch as a whole. When only a few hundred grams or even 1-5 kg of product are taken as a sample of a production batch of several tonnes, this is a daunting prospect and a subject which has given rise to much debate. The ICMSF (ref. 33) discussed the problems of probability and sampling plans with respect to determining the true microbiological status of foods. Since the authors at that time did not have sufficient reliable data regarding the type of frequency distribution of micro-organisms in food, they discounted the use of variables sampling plans and developed attributes plans. The latter plans are relatively independent of the distribution within a batch (or lot) of food. Two types of attribute plan were recommended- a two-class and a three-class. The two-class plan accepts or rejects a batch of food on the basis of the absence of a particular organism from a stated number of samples, e.g. absence of *Salmonella* in five samples of 25 g, or a viable count of micro-organisms not exceeding a stated number. Alternatively, the viable counts of not more than a certain number (c) of the total samples analysed (n) may exceed a stated viable count. This type of plan allows no indeterminate or 'grey' area in whether a batch is accepted or rejected. Three-class plans permit a certain number of the results to lie between an upper (M) and lower limit (m) but none may exceed the upper limit. From the total samples (n) taken and the number (c) permitted to exceed the stated count (m), in both types of plan the probability of acceptance (Pa) of a defective batch of food can be calculated (ref. 33). However, if the microbial population is far from randomly distributed, a two-class plan does not provide as good protection to the consumer. The only way to reduce Pa of a defective batch of food in which non-random distribution occurs is to increase dramatically the number of samples analysed.

Kilsby, Aspinall and Baird-Parker (ref. 34) pointed out that the ICMSF attributes schemes still had a high chance of failing batches of food in which the levels of micro-organisms were well below the safety limits specified. They evolved sampling schemes which were designed to take into account the economic problems of the food producer without increasing the risk to the consumer or increasing the analytical workload. These authors, from published data, showed that the distribution of microbes in many foods is approximately log normal. Thus the statistics based on normal distribution can be used for analysis of data and decision making. Attributes schemes take account only of whether results are above or below the level specified (m or M); the type of scheme based on normal distribution statistics takes into account the actual concentration of micro-organisms in the samples (representing that in the batch) as a continuous variable - a variables sampling plan. The mean log and its standard deviation can be used to describe the level and distribution of microbes in a batch of food if the distribution is log normal; the latter may be determined experimentally by analysis of about thirty samples and showing that there are about as many counts above the mean log as there are below. A chi-square test will be more precise. Utilising the mean log and its standard deviation, the probability of acceptance or rejection of a batch of food with a given percentage of analytical results exceeding a stated value can be calculated for any number of replicate samples.

Kilsby et al. (ref. 34) also suggest using values similar to m and M in the ICMSF attributes scheme, to which they assign an upper critical limit (C) which represents a safety/quality specification, and a lower limit, which represents a value attainable using GMP (Cm). A variables sampling scheme can therefore be used in the same way as the ICMSF three-class sampling plan, but the method gives a specified minimum probability of rejection of a batch if more than a critical proportion of it exceeds the chosen limit, C. Acceptance or rejection will depend solely on the proportion exceeding C. A more complete review and explanation of sampling schemes are given by Kilsby (ref. 35).

Kilsby and Pugh (ref. 36) examined the relevance of the distribution of micro-organisms within batches of food from the aspect of controlling hazards to the consumer. These authors showed that the distribution, assumed to be log-normal, must be taken into account when attempting to define the level of hazard in any particular food-processing operation, a critical control point. If the variance decreases as a result of a processing stage, e.g. mixing, then the mean log increases although the average number of organisms remains constant. When low numbers of organisms are present, this problem is exacerbated, since there is a limit to the sensitivity of the analytical method used. As mixing increases the mean log, this eventually exceeds the detection level in any one sample, thus leading to the erroneous conclusion that the mixing operation is the cause of the contamination detected subsequently.

Sub-lethal Damage

The second problem facing a food microbiologist is that of sub-lethal damage to micro-organisms as a result of the drying process, either from heat or stress imposed by the low A_w. If organisms are not permitted to repair sub-lethal damage before being exposed to selective agents in media, these organisms will not survive and grow and a false impression of the microbiological quality of the dried food will be obtained. Mossel and Corry (ref. 37) reviewed the literature concerned with sub-lethal damage in pathogenic and index bacteria in foods and water. Prior to 1977 many workers had reported marked differences in counts recorded on or in different media, particularly selective and non-selective media. Stresses resulting in sub-lethal damage and sub-optimal recovery on selective media were imposed by freezing and chilling, heating, drying, low pH preservation, and exposure to disinfectants. The difference in counts on selective and non-selective media depends upon the severity and duration of the stress, the type of organism involved and the selective medium, but can vary from one to six log cycles. Furthermore, different batches of the same medium may give rise to markedly different results (ref. 38). These data call into question the results of earlier analyses of dried foods, in which sub-lethal damage was either not recognised or ignored, since samples, after reconstitution, were transferred directly to selective broths or agar media. Even some of the more recent publications may not have fully recognised the problem of sub-lethal damage; for example, Singh _et al._ (ref. 39) enumerated coliforms directly on MacConkey agar. In order to evaluate the efficacy of a recovery medium or technique Mossel and Corry (ref. 37) recommended the construction of a Destruction-Repair curve. Three phases are detected in such a curve; that of an increasing difference in counts obtained on selective and non-selective media as a result of application of the stress, a period following removal of the stress in which there is initially little or no change in the difference in counts followed by a rapidly diminishing difference in counts as resuscitation or repair of sub-lethal damage occurs, and finally the appearance of normal multiplication with little or no difference in counts on the selective and non-selective media. However, in any population of stressed cells, there is likely to be a distribution of stressed cells from those which are only marginally stressed and therefore capable of multiplication almost immediately in or on a selective medium, to those which are severely stressed and therefore require quite a long period (several hours) for resuscitation. This distribution will clearly give rise to problems in enumeration if resuscitation is in a liquid medium; marginally stressed cells will multiply before more severely stressed cells have fully repaired the sub-lethal damage. The most reliable resuscitation method, if enumeration is required, is carried out on a non-selective solid medium followed by replica plating or transfer of a filter membrane to a selective agar medium. If Most Probable Number

(MPN) estimates are sufficient, then resuscitation in liquid medium is satisfactory with subsequent transfer to, or addition to the resuscitation medium of selective agents for the organisms sought. There are conflicting reports on whether a nutritionally 'poor' or 'rich' medium is optimal for resuscitation.

There are two possible forms of sub-lethal damage in dried micro-organisms - that of heat damage and that of dehydration. Conditions for optimum repair of these may differ. Ray, Jezeski and Busta (ref. 40) attempted to take account of dehydration damage by including two levels of rehydration, 1:10 and 1:2.5, dry product:lactose broth, in their study of Salmonella survival in contaminated dried milk products. Two temperatures, 25°C and 45°C, were also used during rehydration for one hour. Although there were no major differences, rehydration at a ratio of 1:2.5 gave marginally higher MPN values. However, in both rehydration ratios, the dry sample was rapidly mixed with lactose broth, undoubtedly raising the water content of damaged cells very rapidly. Ray, Jezeski and Busta (ref. 41) examined these effects in more detail with freeze-dried Salm.anatum. There was little difference between media for resuscitation, from water as a 'poor' medium to a 'rich' medium containing peptones, yeast extract and casamino acids, or lactose broth. Decreasing the A_w of the rehydration medium, by adding lactose, sucrose or milk solids, increased the number of cells recovered compared with rehydration in water. Rehydration temperatures of 15° and 25°C gave initial recoveries higher than at 35° or 45°C, although earlier growth occurred at the latter temperatures. Whilst Harrewijn and Mossel (ref. 6) recognised the need for resuscitation of dried cells, in their methods for microbiological analysis no account was taken of rehydration ratios or for slow (soaking) rehydration. Similarly, Silliker and Gabis (ref. 42) and Gabis and Silliker (ref. 43) took no account of rehydration as a possible problem. Van Schothorst, van Leusden, de Gier, Rijnierse and Veen (ref. 44), using buffered peptone water (BPW) as a non-selective pre-enrichment broth, showed quite conclusively that higher recoveries of Salmonella cells in dried milk were obtained if the powder was allowed to rehydrate slowly on the surface of the medium, or mixed at a low ratio with medium (1:less than 5). The period of reconstitution at a dilution ratio of 1:2 before diluting to 1:9 was not of great importance provided it was not less than 10-20 min in batches of product with a low contamination rate. These authors therefore recommended reconstituting 25 g of milk powder with 50 ml of BPW for 30 min at ambient temperature, followed by dilution with a further 175 ml BPW and incubation at 37°C for 16-20 h. In a collaborative study of pre-enrichment broths for recovery of *Salmonella* from milk chocolate and edible casein, this advice was apparently not heeded (ref. 45). Later, however, Andrews, Wilson and Poelma (ref. 46) extended the observations of van Schothorst *et al.* (ref. 44); markedly improved detection of salmonellae by slow rehydration ('soaking') or by low solid:liquid ratios was achieved, particularly when the level of *Salmonella* contamination was very low, in non-fat dried milk (less than 0.2 cfu/g). In contrast to van Schothorst *et al.* (ref. 44) results, however, the soak method gave marginally better recoveries of salmonellae than 1:2 ratio and swirl method. However, the soak method does not apparently give optimal results with all dried foods. Wilson, Andrews, Poelma and Wagner (ref. 47) found that the 1:9 dilution ratio with swirling gave higher recoveries of salmonellae from brewer's yeast, dried active yeast and onion powder than the soak method. For soya flour, the soak method gave higher *Salmonella* recovery than the swirl method. However, these authors also remarked on some potential problems with the soak method. In some cases the product under test may not be fully wetted after incubation, and any trapped organisms would have been unable to multiply. For acidic samples the lack of pH adjustment in the pre-enrichment phase would enhance the damage to cells. It appears that the optimum conditions for the resuscitation of sub-lethally damaged cells must be determined for each type of dried food.

Inhibitory Effects of Cocoa

A problem specific to the recovery of salmonellae from cocoa powder was highlighted by several workers who demonstrated that cocoa and chocolate exhibited antimicrobial properties. For example, Cathcart and Merz (ref. 48) observed that bakery custard fillings containing cocoa or chocolate inhibited the growth of *Staph.aureus*. Busta and Speck (ref. 49) observed that cocoa powder inhibited the growth of salmonellae in cultures agitated before and/or during incubation, but not in static cultures. Zapatka, Varney and Sinskey (ref. 50) confirmed these results and demonstrated that non-fat dried milk or casein minimised the inhibitory effect. Wilson, Andrews and Poelma (ref. 51), comparing pre-enrichment media for recovery of salmonellae from milk chocolate, and Poelma, Andrews and Wilson (ref. 45) demonstrated that non-fat dried milk containing 0.002% Brilliant Green was marginally superior to BPW for recovery of salmonellae. It is now routine to pre-enrich any cocoa or chocolate product in non-fat dried milk, with or without Brilliant Green.

MICROBIOLOGICAL CRITERIA FOR DRIED FOODS

In a symposium on Dried Foods, Buttiaux and Beerens (ref. 52) dealt with the general principles and various factors governing the hygienic quality of dehydrated foods. These authors considered the factors involved in determining the final microbiological condition of dried foods, such as the nature and number of the initial microbes and their individual reactions to heating and drying, secondary contamination during processing and method of dehydration, and final packing and possibility of contaminating the final product. The multiplicity of these factors pertaining to different foods explains the variety of problems associated with the range of dried foods. The problems to be faced when determining the microbiological quality of a dried food and guidelines drawn up for agreement include: obtaining a representative sample; resuscitation of the sub-lethally damaged cells; germination of spores and selection of those organisms to be sought which are relevant to the health of the consumer or important to the commercial use of the product, e.g. effects of indigenous microbes on the stability of a product in which the dried food may be incorporated. Buttiaux and Beerens (ref. 52) warned that any standards proposed need to be the subject of thorough studies and be relevant to the end-use of the product. Holtzapffel and Coutinho (ref. 54) reported studies on the microbiological condition of vegetables, Krugers, Dagneaux and Mossel (ref. 54) of dried soups and of spices, Meursing and Slot (ref. 55) of cocoa powder and Hauge (ref. 56) of fish meal. From these and other studies, Skovgaard (ref. 57) suggested two general standards for various dried foods. The first category included dried soups, meat, yeast extract and vegetables, and was generally more lax than the category which included dietetic foods, baby foods and certain instantised foods (Table 3). As part of these suggested standards, the media and methods to be employed were also specified, which is a most important point (see previously, re sub-lethal damage). However, in the discussion to Skovgaard's paper (ref. 57), there was obviously some dissent because all dried foods were placed in one of only two categories; specific

TABLE 3

Suggested standards for certain dried foods (ref. 57)

Criterion	Category 1*	Category 2*
Aerobic mesophilic organisms	less than 10^5/g	less than 10^4/g
Clostridia	less than 100/g	less than 10/g
Cl.botulinum	0	0
"Coliforms"	less than 100/g	less than 10/g
E.coli I	less than 10/g	less than 10/g
Enterobacteriaceae	less than 10/g	less than 10/g
Bacillus cereus	less than 10^3/g	less than 10/g
Staphylococci	less than 100/g	less than 10/g
Salmonella spp.	Absent in 50 g	Absent in 50 g

* Category 1 includes dried soups, meat, yeast extract and vegetables.

* Category 2 includes dietetic foods, baby foods and certain instantised foods.

standards should be set for specific foods, it was argued and also, the aspects of industrial control and public health should be kept separate. Van Sprang (ref. 58) also presented bacteriological specifications for dried infant and weaning foods, which were claimed to be medically justified and technologically attainable. In most instances these specifications were more stringent than those for the Category 2 foods of Skovgaard (ref. 57), e.g. Salmonella absent in 100 g, E.coli absent in 10 g. However, a lively discussion again apparently took place with differing points of view. In particular, the importance to be attached to certain types of counts was discussed, together with a recognition that certain individuals, e.g. babies, hospital patients, etc., will be more susceptible to microbial attack. If the criteria proposed by Skovgaard (ref. 57) or van Sprang (ref. 58) were applied to the spray-dried milk powder from the Sakha plant, as reported by Taha, Naguib and Abd-el-Ghani (ref. 4), the product would be unacceptable. Harrewijn and Mossel (ref. 6) advocated a combination of overall quality assurance in which Good Manufacturing Practice (GMP) constituted the major effort supported by such checks - microbiological, chemical, physical, enzymological - as were appropriate to the commodity being dried and the individual operations constituting the conversion of raw materials to final product. In such a system, these authors contended, drawing and examination of samples of final products make sense as it is only the final check on the success of the manufacturing process. Harrewijn and Mossel (ref. 6) also sub-divided dried foods into those pasteurised in the course of manufacture, e.g. milk, eggs, and those exposed to much lower heat treatments or none at all, e.g. flour, and advocated a much less stringent examination of the latter category. Nevertheless, microbiological specification was given for dried foods (ref. 6), which did not differ significantly from that of Skovgaard (ref. 57). Singh _et al_. (ref. 39) reported on the bacteriological quality of infant milk foods in India. The total counts were all 10^3 - less than 10^4/g, staphylococci 10^2 - 10^3/g, _B.cereus_ less than 1 - 10^2/g and coliforms less than 1/g. These authors expressed concern with regard to the high staphylococcal and _B.cereus_ counts and further examined those strains isolated for characteristics associated with food-poisoning strains.

Probably the most widely respected sampling plans and microbiological limits for a wide variety of foods at present are those published by the International Commission on Microbiological Specifications for Foods (ref. 33). The analytical methods to be used for the examination of samples are specified as part of the microbiological criteria. The type of product, the intended consumer and any further processing of the dried food prior to consumption are taken into account when assessing the degree of hazard (the 'Case') and thus deciding the plan class and the microbiological limits. Some examples of the microbiological criteria for dried foods are given in

Table 4. In view of the now recognised risk of salmonellosis from cocoa products, particularly chocolate, most microbiologists in this industry would now also examine 25 g or 50 g for the presence of Salmonella.

In a recent survey of some dried foods (cocoa powder, dry instant chocolate drink mix, dry non-dairy coffee creamer) by Payne et al. (ref. 59) in the USA, some 1,300-1,500 samples of each product were analysed

TABLE 4

Microbiological criteria for some dried foods and ingredients (ref. 34)

Product	Test	Case	Plan	n	c	Limit/g m	Limit/g M
Egg products	SPC*	2	3	5	2	10^4	10^6
	Coliforms or Enterobacteriaceae	5	3	5	2	10	10^3
	Salmonella	11	2	10	0	0	–
Special dietary foods	SPC	3	3	5	1	10^4	10^6
	E.coli	5	3	5	2	less than 3	10
	Staph.aureus	9	3	10	1	10	10^2
	B.cereus	9	3	10	1	10^2	10^4
	Cl.perfringens	9	3	10	1	10^2	10^3
	Salmonella	15	2	60	0	0	–
Cocoa	SPC	2	3	5	2	10^4	10^6
	Moulds	2	3	5	2	10^2	10^4
Coconut	Moulds	2	3	5	2	10^2	10^4
	Coliforms or Enterobacteriaceae	5	3	5	2	10	10^3
	Salmonella	15	2	60	0	0	–
Dried milk	SPC	5	3	5	2	5×10^4	5×10^5
	Coliforms	5	3	5	2	less than 3	10^2
	Staph.aureus	8	3	5	1	10	10^2
Dried milk for relief work	Staph.aureus	9	3	10	1	10	10^2
	Salmonella	12	2	60	0	0	–

* SPC, Standard Plate Count using pour plates of Plate Count Agar, incubated at 35°C for 48 h.

for aerobic plate count, coliforms, E.coli and Staph.aureus. On the basis of the sampling plans and microbial limits of ICMSF (ref. 33), none of the cocoa samples was defective, although 31% were of marginal quality, i.e. exceeding m but less than M. Collins-Thompson, Weiss, Riedel and Charbonneau (ref. 60) had recommended a slightly different three-class plan for cocoa on the basis of a survey in Canada (n = 3, c = 1, m = 10^5/g, M = 10^6/g for aerobic plate counts and n = 3, c = 1, m = 1.8/g, M = 10/g for confirmed coliforms). Using this sampling plan and microbiological criteria, only one unit of the cocoa powder sampled would be of marginal quality by the aerobic colony count but 5.7% of units would be marginal and 0.7% defective if the coliform criteria were applied.

Good Manufacturing Practice

However stringent a sampling plan and microbiological criteria applied to dried foods, there is still only a statistical probability of detecting a defective batch of product. Much of this arises from the difficulties of obtaining truly representative samples of very large batches of product. Mossel (refs. 61, 62) argued strongly that safety and keeping quality of foods are best maintained by factory surveillance using a range of tests appropriate to the product, rather than the "analytical" approach in which final product samples are examined and action taken retrospectively. This approach was further developed and expanded by Mossel to include "longitudinally integrated processing for safety (LIPS)", which incorporated the view that prevention is better than cure but that both aspects should be employed (ref. 63). In practice, steps should be taken in designing and operating a process to prevent contamination and minimise conditions in which microbes can multiply and to "cure" any possible problem by terminal processing for safety. However, as Mossel (ref. 63) comments, proper design of processes relies on risk assessment, which is relatively new to food microbiology, and requires data on which to base a relevant analysis of the risk to product and consumer. Monitoring of final product is useful in assessing whether GMP has been attained in the manufacturing of that product, but will give little assistance in determining the efficacy of each operation, and therefore in attaining GMP.

REFERENCES

1 E.L. Crossley and A.T.R. Mattick in XII Int. Dairy Congress, Stockholm, Vol. 3, Sect. 3B; 1949, pp.449-453.
2 E.M. Foster, F.E. Nelson, M.L. Speck, R.N. Doetsch and J.C. Olson (Jr.) (Eds.), Dairy Microbiology, Prentice-Hall, Englewood Cliffs, New Jersey, U.S.A., 1958.
3 B.P. Keogh, J. Dairy Res. 38(1971)91-111.
4 S.M. Taha, K.H. Naguib and S. Abd-el-Ghani, J. Milk Food Technol. 35(1972)35-38.
5 K.H. Naguib, S. Abd-el-Ghani and S.M. Taha, J. Milk Food Technol. 35(1972)45-48.
6 G.A. Harrewijn and D.A.A. Mossel in J.S. Furtado (Ed.), Global Impacts of Applied Microbiology, 4th Int. Symp., Sociedade Brasileira de Microbiologia, Brazil 1974, pp.601-613.
7 S.S. Thompson, L.G. Harmon and C.M. Stine, J. Food Prot. 41(1978)16-19.
8 R.N. Collins, M.D. Treger, J.B. Goldsby, J.R. Boring, D.B. Coohon and R.N. Barr, J. Am. Med. Assoc. 203(1968)118-124.
9 F.E. McDonough and R.E. Hargrove, J. Dairy Sci. 51(1968)1587-1591.
10 T.E. Galesloot and J. Stadhouders, in E.H. Kampelmacher, M. Ingram and D.A.A. Mossel (Eds.), The Microbiology of Dried Foods: Proc. Sixth Int. Symp. on Food Microbiology, Grafische Industrie, Haarlem, The Netherlands, 1969 pp.341-343.
11 J.J.LiCari and N.N. Potter, J. Dairy Sci. 53(1970)871-876.
12 D.L. Miller, J.M. Goepfert and C.H. Amundson, J. Food Sci. 37(1972)828-831.
13 F. Espina and V.S. Packard, J. Food Prot. 42(1979)149-152.
14 A. Chopin, G. Mocquot and Y. LeGraet, Can. J. Microbiol. 23(1977)755-762.
15 A.L.H. Daemen, Neth. Milk and Dairy J. 35(1981)133-144.
16 A.L.H. Daemen and H.J. van der Stege, Neth. Milk and Dairy J. 36(1982)211-219.
17 A.L.H. Daemen, A. Kruk and H.J. van der Stege, Neth. Milk and Dairy J. 37(1983)213-228.
18 J.E.L. Corry, in R. Davies, G.G. Birch and K.J. Parker (Eds.), Intermediate Moisture Foods, Academic Press, London and New York, 1976 pp.215-238.
19 R. Pawsey and R. Davies in R. Davies, G.G. Birch and K.J. Parker (Eds.), Intermediate Moisture Foods, Academic Press, London and New York, 1976 pp.182-202.
20 A. Anellis, J. Lubas and M.M. Rayman, Food Res. 19(1954)377-395.
21 R.P. Elliott and B.C. Hobbs in J.H. Silliker, R.P. Elliott, A.C. Baird-Parker, F.L. Bryan, J.H.B. Christian, D.S. Clark, J.C. Olson (Jr.) and T.A. Roberts (Eds.), Microbial Ecology of Foods Vol. II. Food Commodities, I.C.M.S.F., Academic Press, London and New York, Ch.19, pp.521-566.
22 W.H. Lee, C.L. Staples and J.C. Olson (Jr.), J. Food Sci. 40(1975)119-120.
23 D.E. Walsh and B.R. Funke, J. Food Sci. 40(1975)714-716.
24 D.E. Walsh, B.R. Funke and K.R. Graalum, J. Food Sci. 39(1974)1105-1106.
25 F.H. Hsieh, K. Acott and T.P. Labuza, J. Milk and Food Technol. 39(1976)619-623.

26 W. Seiler and W. Bruhn, Dragoco Rep. 2(1984)35-44.
27 E.M. Foster, Candy Ind. and Confect. J. 131(1968)63-73.
28 E.M. Foster, Manuf. Confect. 48(1968)33-37.
29 J.M. Goepfert and R.A. Biggie, Appl. Microbiol. 16(1968)1939-1940.
30 J.C. Barrile and J.F. Cone, Appl. Microbiol. 19(1970)177-178.
31 P.C. Craven, D.C. Mackel, W.B. Baine, W.H. Barker, E.J. Gangarosa, M. Goldfield, H. Rosenfeld, R. Altman, G. Lachapelle, J.W. Davies and R.C. Swanson, The Lancet (i)(1975) 788-793.
32 O.N. Gill, P.N. Sockett, C.L.R. Bartlett, M.S.B. Vaile, B. Rowe, R.J. Gilbert, C. Dulake, H.C. Murrell and S. Salmaso, The Lancet (i)(1983)574-577.
33 I.C.M.S.F., in M. Ingram, D.F. Bray, D.S. Clark, C.E. Dolman, R.P. Elliott and F.S. Thatcher (Eds.), Micro-organisms in Foods 2. Sampling for microbiological analysis: Principles and specific applications, Univ. Toronto Press 1978.
34 D.C. Kilsby, L.J. Aspinall and A.C. Baird-Parker, J. Appl. Bact. 46(1979)591-599.
35 D.C. Kibby in M.H. Brown (Ed.), Meat Microbiology, Applied Science Publishers Ltd., London and New York 1982, Ch. 10, 387-421.
36 D.C. Kilsby and M.E. Pugh, J. Appl. Bact. 51(1981)345-354.
37 D.A.A. Mossel and J.E.L. Corry, Alimenta-Sonderausgabe 16(1977)19-34.
38 R.B. Read and A.L. Reyes, Appl. Microbiol 16(1968)746-748.
39 R.S. Singh, S. Singh, V.K. Batish and B. Ranganathan, J. Food Prot. 43(1980)340-342.
40 B. Ray, J.J. Jezeski and F.F. Busta, J. Milk Food Technol. 34(1971)423-427.
41 B. Ray, J.J. Jeseski and F.F. Busta, Appl. Microbiol. 22(1971)184-189.
42 J.H. Silliker and D.A. Gabis, Can. J. Microbiol. 22(1976)971-974.
43 D.A. Gabis and J.H. Silliker, Can. J. Microbiol. 23(1977)1225-1231.
44 M. van Schothorst, F.M. van Leusden, E. de Gier, V.F.M. Rijnierse and A.J.D. Veen, J. Food Prot. 42(1979)936-937.
45 P.L. Poelma, W.H. Andrews and C.R. Wilson, J. Assoc. Off. Anal. Chem. 64(1981)893-898.
46 W.H. Andrews, C.R. Wilson and P.L. Poelma, J. Food Sci. 48(1983)1162-1165.
47 C.R. Wilson, W.H. Andrews, P.L. Poelma and D.E. Wagner, J. Food Prot. 48(1985)505-508.
48 W.H. Cathcart and A. Merz, Food Res. 17(1942)96-99.
49 F.F. Busta and M.L. Speck, Appl. Microbiol. 16(1968)424-425.
50 F.A. Zapatka, G.W. Varney and A.J. Sinskey, J. Appl. Bact. 42(1977)21-25.
51 C.R. Wilson, W.H. Andrews and P.L. Poelma, J. Food Sci. 45(1980)310-316.
52 R. Buttiaux and H. Beerens in E.H. Kampelmacher, M. Ingram and D.A.A. Mossel (Eds.), The Microbiology of Dried Foods, Proc. Sixth Int. Symp. on Food Microbiology, Int. Assoc. Microbiol. Socs., Grafische Industrie, Haarlem, The Netherlands, 1969, pp.48-66.
53 D. Holtzapffel and H.E. Coutinho in E.H. Kampelmacher, M. Ingram and D.A.A. Mossel (Eds.) The Microbiology of Dried Foods, Proc. Sixth Int. Symp. on Food Microbiology, Int. Assoc. Microbiol. Socs., Grafische Industrie, Haarlem, The Netherlands, 1969, pp.387-410.

54 E.L. Krugers Dagneaux and D.A.A. Mossel in E.H. Kampelmachar, M. Ingram and D.A.A. Mossel (Eds.), The Microbiology of Dried Foods, Proc. Sixth Int. Symp. on Food Microbiology, Int. Assoc. Microbiol. Socs., Grafische Industrie, Haarlem, The Netherlands, 1969, pp.411-432.
55 E.H. Meursing and H. Slot in E.H. Kampelmacher, M. Ingram and D.A.A Mossel (Eds.), The Microbiology of Dried Foods, Proc. Sixth Int. Symp. on Food Microbiology, Int. Assoc. Microbiol. Socs., Grafische Industrie, Haarlem, The Netherlands, 1969, pp.433-445.
56 S. Hauge in E.H. Kampelmacher, M. Ingram and D.A.A. Mossel (Eds.), The Microbiology of Dried Foods, Proc. Sixth Int. Symp. on Food Microbiology, Int. Assoc. Microbiol. Socs., Grafische Industrie, Haarlem, The Netherlands, 1969, pp.446-454.
57 N. Skovgaard, in E.H. Kampelmacher, M. Ingram and D.A.A. Mossel (Eds.), The Microbiology of Dried Foods, Proc. Sixth Int. Symp. on Food Microbiology, Int. Assoc. Microbiol. Socs., Grafische Industrie, Haarlem, The Netherlands, 1969, pp.457-479.
58 F.J. van Sprang in E.H. Kampelmacher, M. Ingram and D.A.A. Mossel (Eds.), The Microbiology of Dried Foods, Proc. Sixth Int. Symp. on Food Microbiology, Int. Assoc. Microbiol. Socs., Grafische Industrie, Haarlem, The Netherlands, 1969, pp.496-510.
59 W.L. Payne, A.P. Duran, J.M. Lanier, A.H. Schwab, R.B. Read (Jr.), B.A. Wentz and R.J. Barnard, J. Food Prot. 46(1983)733-736.
60 D.L. Collins-Thompson, K.F. Weiss, G.W. Riedel and S. Charbonneau, Can. Inst. Food Sci. Technol. J. 11(1978)177-179.
61 D.A.A. Mossel, J. Milk Food Technol. 32(1969)155-171.
62 D.A.A. Mossel, J. Food Technol. 3(1969)401-421.
63 D.A.A. Mossel, Microbiology of Foods, The University of Utrecht, Faculty of Veterinary Medicine, 1982.

EVAPORATOR DESIGN

G. HAHN

Gea-Wiegand GmbH, Karlsruhe, West-Germany

ABSTRACT

A short review of the history of evaporation is provided to illustrate how far evaporation technology has advanced. An evaporation plant consists of a group of essential components, which will be presented in sequence. Their function and design criteria will be discussed. The successful operation of an evaporation plant depends on the correct design of its components and their coordinated trouble-free interaction.

INTRODUCTION

A modern evaporation plant consists of many individual components, which are essential for the operation of the plant, (Fig. 1). Trouble-free operation can only be guaranteed if these components are correctly designed and coordinated. Different types of equipment will be compared, their operation explained, and the criteria for their design will be discussed.

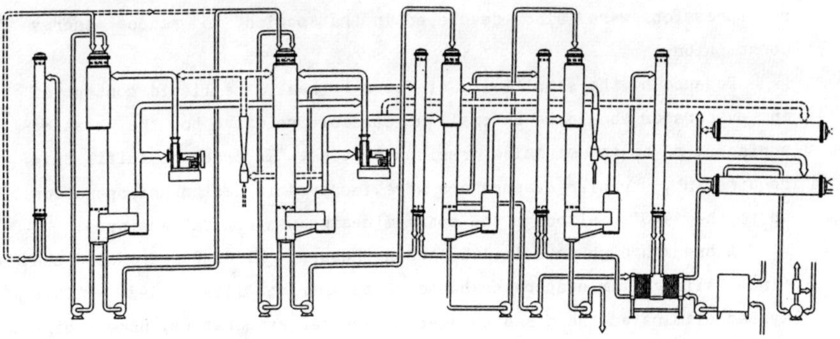

Fig. 1: Four-effect falling-film evaporation plant with two high-capacity fans and steam jet vapour recompressors.

REVIEW

Before the design of modern evaporation plants is examined, the history of evaporation technology will be reviewed, so that progress towards modern economic evaporation plants can be better understood.

At the beginning of the industrialization of food technology, batch processes were widely used. In batch evaporation, the process liquid was boiled in a vessel until enough water was evaporated and the product had the required consistency. Modern evaporation plants are far removed from this basic idea.

The capacity of early evaporation pans was increased by the use of heating coils or by the use of tubular heat exchangers to increase increase heating surface area. Many undesirable changes were caused in the product during evaporation at temperatures up to 100°C and at atmospheric pressure. Continuous vacuum evaporators were then developed.

The first continuous multiple-effect evaporators were used in the middle of the last century in the sugar industry. Before about 1910 however, most evaporation plants had only one effect. In general, the evaporation capacity did not exceed 300 to 1500 kg/h water evaporation due to the low throughputs required. In the 1920's and 1930's double-effect circulation evaporators dominated the market. Certain advances were made in the use of new materials. Stainless steel was increasingly used instead of copper and aluminium, which greatly improved cleanability. The first plants with steam jet vapour recompression were also developed in that period to reduce energy consumption.

Because of the sensitivity of foods to heat, the liquid content of an evaporator should be as small as possible relative to the heating surfaces, so as to minimise residence time. In order to fulfil this requirement, tubular evaporators were improved in design and operation up to the 1940's, although the optimum design was not yet achieved.

A breakthrough came in 1953, when Messrs. WIEGAND introduced the first falling-film evaporator based on an American idea of 1935. This system eliminated the disadvantages of earlier evaporators, namely high temperature differences, long residence times, short service life, limited number of effects, and low final concentration.

Within a few years the falling-film evaporator had replaced nearly all other evaporator types. Development continued, and in 1966 the first quadruple-effect evaporation plant was produced, in 1974 the first five-effect plant, in 1976 the first six-effect plant, and shortly after this, the first seven-effect evaporator was installed in the food industry. In 1978 mechanical vapour recompression was introduced on a triple-effect falling-film evaporation plant in the milk industry.

Because of the great importance of the falling-film evaporation plant in the food industry, only this type will now be discussed.

COMPONENTS OF AN EVAPORATION PLANT

If one considers the flow sheet of a modern, multi-effect evaporation plant, one will see that apart from the actual calandrias, many other components are necessary to operate the plant. The trouble-free operation of an evaporation plant depends on the correct design of these components and on their coordinated operation. These aspects will now be discussed.

PREHEATING

After feeding the product into the evaporation plant via a float balance tank and feed pump, the first process step is to preheat the product. In general, liquid food is stored at between 5 and 20°C for quality reasons.

Unless the product has been heated in another process step, it must be pre-heated to at least the boiling temperature of the first evaporation effect, so that evaporation can start immediately the product enters the heating tubes. The most common boiling temperatures of the first effect in the food industry are between 40 and 110 deg. C. If the product is below its boiling point, it has to be preheated to boiling point in the heating tube before concentration begins. This preheating in the heating tube is achieved mainly by direct condensation of vapours from the lower part of the heating tube, and thus back flow may occur, which can destabilize the operation of the evaporator.

When the preheaters are designed, it must be taken into account that heat transfer surface area and capital costs are reduced by increasing the temperature difference between product and heating medium. At the same time the residence time is reduced. On the other hand, larger temperature differences can cause undesirable changes in temperature-sensitive products and deposits may be formed on the heat transfer surfaces.

Modern evaporation plants are operated with small temperature differences between the individual effects for reasons of thermal economy. Hence, preheaters will also operate on small temperature differences, with correspondingly large heat transfer surfaces. As a consequence, the design has to achieve a balance between short residence time and financial advantages of a small heat transfer surface on the one hand, and low operating costs and long service life on the other.

The residence time is determined not only by the area of the heat transfer surface, but also by the type of preheater. Another important factor is the relationship between heat transfer surface and the liquid content of the preheater tube. By decreasing tube diameter, residence time is shortened, but the smaller the tube diameter, the greater the pressure loss in the preheater. Therefore tube diameters are usually between 25 and 70 mm, in order to achieve the optimum balance between pressure drop and liquid volume.

Three types of preheaters are used:-

- plate heat exchangers
- spiral tube heat exchangers
- straight tube heat exchangers

The plate heat exchanger has a high heat transfer coefficient , very low liquid content and average pressure losses, but is suitable only for heat exchange between liquids. It is used mainly for the preheating of product with condensate, concentrate, or warm water produced as waste heat from internal combustion engines.

If there are solid particles in the product, a plate heat exchanger should not be used, as the particles may block the narrow product passages. The same problem may also occur with protein-containing products, due to the formation of deposits if the preheating temperature lies above the denaturation temperature. Hence a plate heat exchanger is usually used in the first step of the preheating process.

The preheater most frequently employed in evaporation plants is the spiral type. Due to the secondary flows which develop in the spiral, the spiral tube has better heat transfer coefficients than straight tubes and a reduced tendency to foul. It can be manufactured at a moderate price, as it is installed directly into the calandria or the condenser, and thus no separate heating jacket or pipework for vapour and condensate are required. The disadvantages, however, are that inspection of the tube is impossible, and replacement is difficult.

For feed rates of more than 10 ton/h, two tube coils are usually installed in parallel, in order to reduce the residence time. Due to difficulties in production, only tube spirals with nominal diameters between DN 400 and DN 1500 can be manufactured. Hence straight tube preheaters are employed for very small and very large calandrias. This preheater type has the advantage that the tubes can be inspected when the covers are removed. It is, however, more expensive to produce and install than the spiral tube type. In addition, straight tubes are more susceptible to clogging. Three to ten tubes are usually arranged in parallel, and this arrangement has the advantage of small pressure losses and short residence times.

Straight tube preheaters are especially useful as intermediate preheaters where product of high concentration is being heated, and where equipment inspection is essential.

HEATING AND HOLDING

Following preheating, further heating of the product may be required to a temperature exceeding the boiling temperature of the first effect, for the purpose of enzyme-inactivation, pasteurisation, sterilization, heat stabilisation or heat classification of the finished powder.

Two methods of product heating are used:-

- indirect heating
- direct heating

In case of indirect heating, heat is transferred to the product by condensing steam on the external tube wall (Fig. 2).

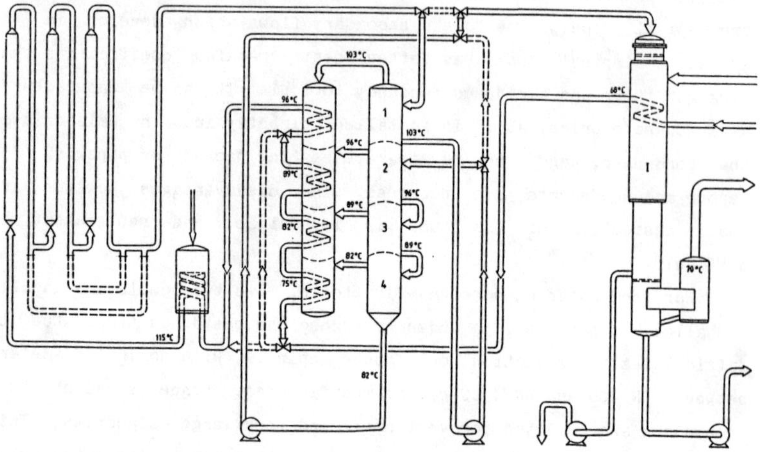

Fig. 2: Heating system with flashing

The same design criteria apply to heaters as to preheaters, in regard to temperature difference, liquid content, tube diameter, residence time, and cost. Since a relatively large heating effect is required during this heating process, special account must be taken of the residence time in order to minimize product damage. Experience has shown that the straight tube preheater is more suitable for indirect heating than the spiral tube. Because a running time of at least 20 hours is required, the plate heat exchanger can not be used.

Product heaters for food can operate effectively at temperatures up to 130°C, provided that the heat transfer surfaces and the product velocity are calculated correctly and the heating steam is saturated not superheated. A temperature difference of about 3-4°C makes it possible to achieve the required operating time of at least 20 hours.

In direct heating, the heating steam and product are mixed, giving very short heating times and gentle product treatment. Because the product is diluted by steam, the evaporator has to be designed for increased capacity. Furthermore, the production of culinary-grade steam is not a straightforward matter.

Direct heating may be done by

- Direct Steam Injection (DSI)
- Mix heater

The direct steam injection process can be installed at reasonable cost, it is space-saving, and the heating temperature is variable over a wide range (Fig. 3).

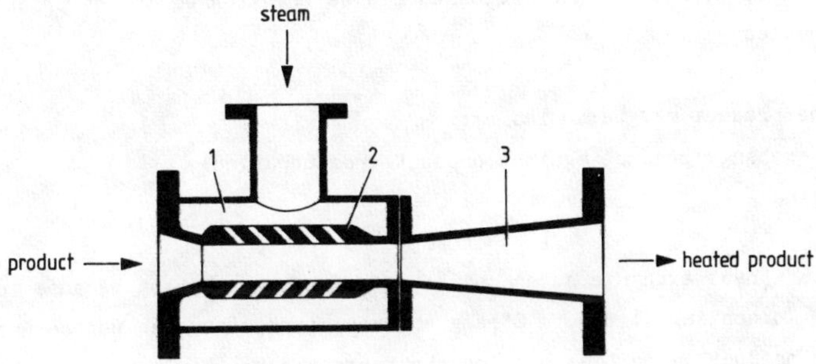

Fig. 3: DSI-unit (1-casing, 2-condensation nozzles, 3-diffuser)

For the trouble-free operation of this equipment, it is essential that steam pressure and product pressure are harmonized. If the steam pressure is too low, product may enter the steam chamber and clog the steam entry holes. If the steam pressure is too high, condensation of the steam in the product may be prevented and the product will be burned.

Overheating of the product can be prevented if the product is sprayed into vapour at the appropriate temperature in a mix heater. However, difficulties may arise in the case of products which tend to foam.

Despite the theoretical advantages of short residence times and careful product treatment offered by the direct heating processes, indirect heating systems prevail on the market, being more stable and more reliable. The main task in the design of the heating system is to find the optimum process for each product and each heating temperature, to ensure trouble-free operation, careful product heating and long operating times.

If the thermal effect from heating-up is not sufficient, a holding section may be arranged downstream of the product heater. In order to achieve certain product properties and to stabilize proteins before further heating, the holding phase can also take place during heating, for instance, for the evaporation of condensed milk.

FLASHING

Before entering the first evaporation effect, the heated product is cooled down, and the heat released can be transferred to the product to be heated.

The product may be cooled by:

- heat exchange product against product (regeneration)
- flashing

Plate heat exchangers can not be used for regeneration because of reduced operating times. Straight tube preheaters can not be used because product velocities outside the tubes are too low. Hence double tube heat exchangers in which the flowrates of both streams have been carefully calculated are the most suitable.

However, it is more usual to cool the product by flashing, which causes a rapid fall in product temperature. The flash vapour can be used to heat incoming product, either by direct heat exchange in the mix preheater or by indirect heat exchange in the spiral or straight tube preheater. Indirect preheating is preferable because cross-contamination can occur with direct heating.

CALANDRIA

As already mentioned, this discussion is limited to falling-film evaporators, as they are the most widely used in the food industry.

A falling-film evaporator consists of the following main components:

- vertical heating tube bundle with surrounding heating jacket
- calandria upper section with product distribution device
- calandria base section with separator (Fig. 4).

The liquid to be evaporated is pumped into the upper part of the calandria, where it is distributed evenly on the tube sheet, from which it flows down the interior walls of the vertical tubes as a thin boiling film. The downward flow of the film is assisted by the drag force of the moving vapour.

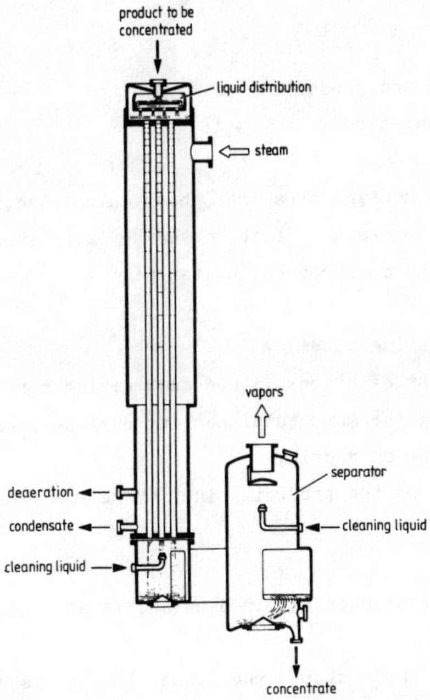

Fig. 4: Falling-film evaporator

The calandria should fulfil the following functions:

- the product should be concentrated equally in all heating tubes
- the product should be concentrated with minimum loss of quality
- during the operation the evaporation capacity should not be significantly reduced.
- the heating tubes should not have dry spots

The most important factor in the trouble-free operation of an evaporator is the liquid coverage of the heating tubes, i.e. the liquid flowrate required to cover each heating tube. Without enough liquid, the liquid film will break near the bottom of the tube, or may not even be formed at the top. Dry spots develop on the tube, the heating surface becomes coated with a deposit, and eventually the tube may get clogged with burned-on product.

The minimum liquid coverage necessary for the safe operation of the evaporator depends on:

- type of product
- concentration of the product
- required operating time

When the calandria heating area has been calculated, the next step is to evaluate tube coverage. If the coverage falls short of a minimum value, the following measures can be taken:

- use of longer heating tubes;
  thus the number of tubes is reduced, giving more liquid per tube while maintaining the same total heating surface area.
- subdivision of the calandria;
  with more passes on the product side there is greater coverage for each tube.
- re-circulation;
  re-circulation of product within a calandria also improves coverage.

In practice, as a first step, the tube length is chosen as long as possible to ensure sufficient coverage. Then the number of passes within the calandria is increased if necessary. Re-circulation should be avoided as far as possible, as it increases the residence time.

In addition to tube coverage, the velocity of the two-phase flow is also of great importance. High velocities cause high pressure losses in the heating tube, increase the boiling temperature, and reduce the effective temperature difference. Moreover, the product and vapour are sprayed off the tube end at high velocity, making the separation of vapour and product more difficult. Thus short tubes with large diameters should be employed in order to minimize pressure loss. On the other hand this reduces liquid coverage of tubes. Hence heating tube selection is a compromise between adequate coverage and pressure drop. Experience has shown that the best solution is to use tubes between eight and twelve meters in length, with a diameter between 30 and 50 mm.

The product distribution in the upper part of the calandria has to fulfil the following functions:

- uniform distribution of the product between heating tubes
- even distribution of the product on the circumference of each heating tube
- effective operation for the duration of a run.

Two different operating principles are used:

- dynamic distribution
- static distribution

An example of dynamic distribution is a full-cone nozzle (Fig. 5), in which the kinetic energy of the liquid is used to atomise the fluid into droplets and distribute them on the upper tube sheet. This distribution device is simple and economic but does not guarantee even distribution, as the droplet pattern of the spray cone may not be uniform. To operate effecitively, the nozzle requires steady feed pressure and feed flowrate and does not function well with varying feed conditions.

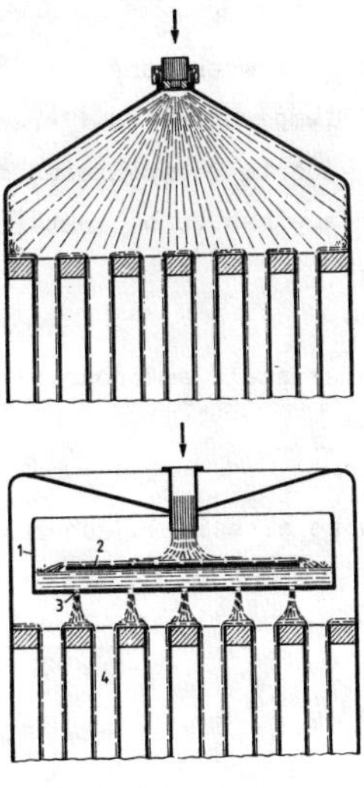

Fig. 5: Distribution devices (1-bowl, 2-bowl hole, 3-smoothing plate, 4-heating tube)

For these reasons, static distribution devices are generally preferred. Even distribution is achieved by having uniform liquid static pressure over all openings in the base of the bowl, giving uniform wetting of all tubes. The system adapts readily to varying feed flowrates by varying the liquid level. Its operation however is sensitive to product contamination, and for products containing solid particles for example fruit juice and coffee extract, the over-flow weirs are slotted.

In designing distribution systems it is important to size the distribution holes so that a head of liquid is created above the smoothing plate. If this is not done, the bowl will either overflow, or be emptied and starve the tubes of liquid.

SEPARATOR

Separators are used to separate the mixture of vapour and concentrated product produced in the calandria.

A separator should fulfil the following requirements:

- High separation efficiency;
  The vapour should as far as possible be free from entrained droplets, so that the condensate of the subsequent calandria or condenser is free from product.
- Small pressure loss;
  If pressure loss is high, the condensation temperature of the subsequent effect is reduced, so that the temperature difference required for concentration is also reduced, and to compensate for pressure and capacity loss extra heating surface has to be installed in the next effect.
- Short residence time;
  The liquid content of the separator should be as small as possible, to minimize product residence time.

The many different separators in use can be classified according to three operating principles:

- gravity separator
- baffle separator
- centrifugal separator

The gravity type separates the free-falling droplets from the vapour by gravity only (Fig. 6). This system requires very large separators, as the vapour velocity has to be reduced until free fall of the droplets occurs.

When the liquid-vapour mixture rebounds from a vertical plate, the system is called a baffle separator. Because of its low density, the vapour may be easily directed to another pass, and the droplets are separated by the impact. This separator produces a satisfactory separation, and is used where the condensate is already contaminated by product components such as volatile acids, and is no longer suitable for use as boiler feed water.

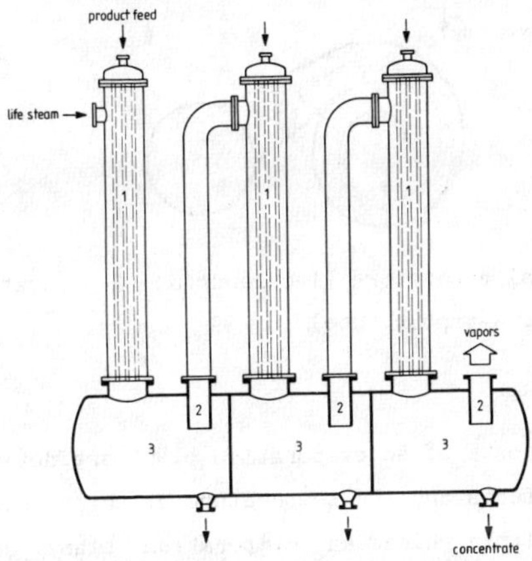

Fig. 6: Gravity separator (1-calandria, 2-vapour tube, 3-separator).

The best separation is produced by the centrifugal separator, (Fig. 7). The vapour-droplet mixture is led into the separator tangentially. The mixture spirals upward, and leaves the separator again through the immersion tube. The acceleration of the flow into the tube causes the centrifugal separation of the droplets.

Pressure loss in the separator is mainly a consequence of the acceleration of the vapour into the immersed vapour tube. Rapid acceleration is necessary to achieve a satisfactory separation effect, but this increases the pressure loss so that part of the temperature difference is lost for the following effect. This can be compensated for by increased heating surface, and it becomes a question of finding the optimum compromise between adequate separation and the higher cost of a larger calandria.

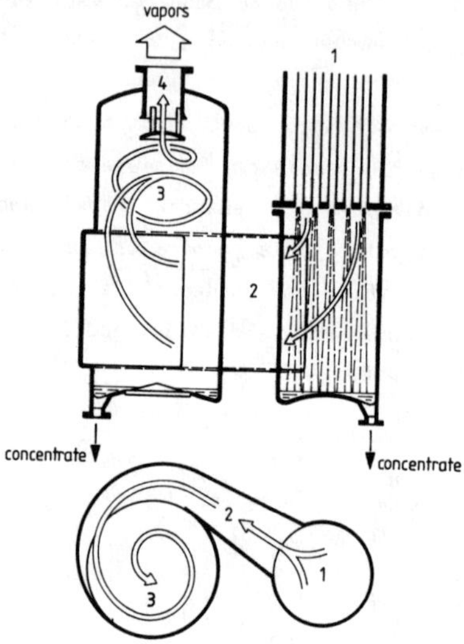

Fig. 7: Centrifugal separator (1-calandria, 2-separator channel, 3-separator, 4-vapour tube).

HEATING

The operating costs of an evaporation plant are determined chiefly by the number of effects and the temperature at which they operate. The temperature range within which an evaporation plant can work in the food industry is limited. The maximum temperature is about 120 C because of the possibility of product damage. In the case of temperature-sensitive and protein-rich products, the maximum boiling temperature is between 60 and 70°C. The minimum temperature is fixed by the condensation temperature of about 40°C which is determined by the cooling water temperature. Within this temperature range between one and seven effects can be used.

The following heating methods are used:

- direct heating
- thermal vapour recompression
- mechanical vapour recompression

Many combinations of these may be employed, ranging from a single-effect directly heated plant with a specific energy consumption of 110%, to a multiple-effect plant with mechanical vapour recompression and a specific energy consumption of 3%. Capital costs also vary widely. Calculation of the minimum overall cost should take into consideration all relevant factors such as operating time, energy cost, and capital costs. In addition, some basic guidelines are available to facilitate heating system selection for evaporation plants.

For small-capacity evaporators with low steam consumption, direct heating is normally used. Direct heating is also recommended where waste steam from a turbine or other source or a steam-air mixture from a drier is available. Directly-heated plants are flexible in regard to temperature-range and product throughput, and do not use a compressor which may have a limited useful operating range.

Only with direct heating is it possible to recover steam condensate separately from product condensate. This is particularly important where product condensate is contaminated with volatiles which render it unsuitable for use as boiler feed.

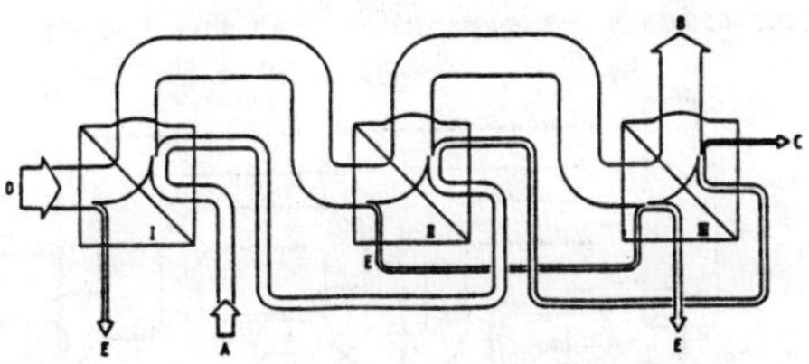

Fig. 8: Heat flow diagram of a triple effect directly heated evaporation plant (A-product, B-residual vapour, C-concentrate, D-heating steam, E-condensate).

Where steam costs are moderate and electricity costs comparatively high, a thermal vapour recompressor is recommended. This simple and relatively inexpensive item has a long operating life, and requires little maintenance. Its economy increases with increasing motive steam pressure and decreasing compression ratio. The thermal vapour recompressor can compress only part of the vapour and the remainder is used in subsequent effects. Just like any other pump the thermal vapour recompressor has a characteristic operating curve, with only one optimum operating point. This restricted flexibility causes high energy consumption if plant operating conditions deviate from design specifications.

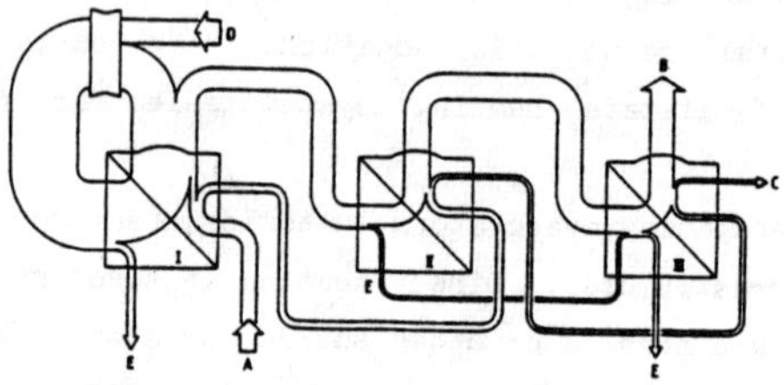

Fig. 9: Heat flow diagram of a triple-effect evaporation plant with thermal vapour recompression (A-product, B-residual vapour, C-concentrate, D-heating steam, E-condensate).

Evaporation plants with mechanical vapour recompression have by far the lowest operating costs (Fig. 10). On the other hand they have the highest capital costs because of the large heating surface area required and the high cost of the mechanical compressor, which is approximately ten times more expensive than the thermal type.

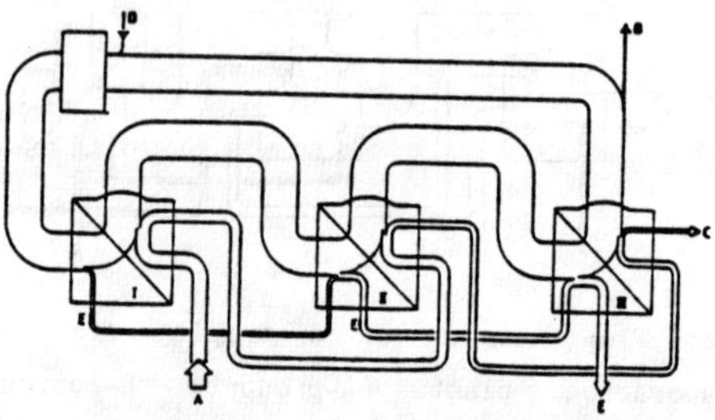

Fig. 10: Heat flow diagram of a triple-effect evaporation plant with mechanical vapour recompression (A-product, B-residual vapour, C-concentrate, D-heating steam, E-condensate).

All the vapour is recirculated by the mechanical vapour recompressor for heating duty, and only a small amount of make-up steam is required. Some cooling water is also used to stabilize plant conditions.

Mechanical recompressors are complicated in design, and require additional expenditure on ancillary equipment such as oil supply and cooling. They also need careful maintenance and supervision of mechanical vibration.

Mechanical recompressors driven by combustion engines are especially effective, as the waste heat of cooling water and exhaust gas can be used for preheating the product, achieving even higher overall efficiency than with electric motor drive. Steam turbine drive is also feasible, and exhaust steam can be used to heat the finisher.

Mechanical vapour recompressors have more or less the same limited flexibility as thermal recompressors. This applies particularly to the high speed radial turbo compressor, which reaches its surge limit if the vapour flow is too small, thus rendering the system unstable.

In conclusion, the choice of evaporator heating system is usually based on energy costs, but other important factors such as equipment reliability and maintenance requirements should also be taken into account.

CONDENSATION AND VACUUM

Vapours produced in evaporation are re-used where possible, either to heat other effects or for product preheating. Surplus vapour must then be condensed and the heat of condensation removed from the system.

The tasks of a condenser are:

- to maintain the heat balance of an evaporation plant
- to stabilize the temperature in each effect
- to produce and maintain vacuum

Condenser cooling water is normally provided from a cooling tower, although in certain cases it may be extracted from rivers, lakes or the sea.

Most condensers are either water-cooled surface condensers or mixing condensers.

In the surface condenser (Fig. 11) the cooling water is completely separated from the vapour being condensed, in contrast to the mixing condenser. The mixing condenser (Fig. 12) requires less cooling water, because the outlet temperature of the cooling water is higher for a given difference between condensation temperature and cooling water temperature. In addition, the mixing condenser is simple in design, is resistant to contamination, and moderate in cost.

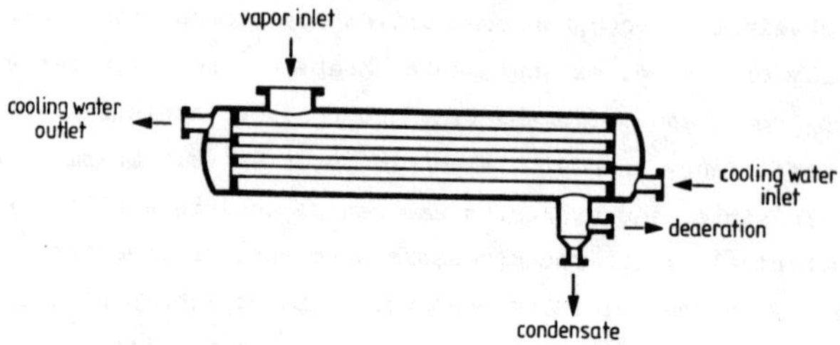

Fig. 11 Surface condenser.

It is possible however that product particles from the separator of the last effect may reach the mixing condenser and contaminate the cooling water. If the water is cooled in a cooling tower, fouling may occur, and the product could be contaminated if contaminated cooling water got into the last separator due to faulty evaporator operation. For this reason, surface condensers are the norm in modern evaporators.

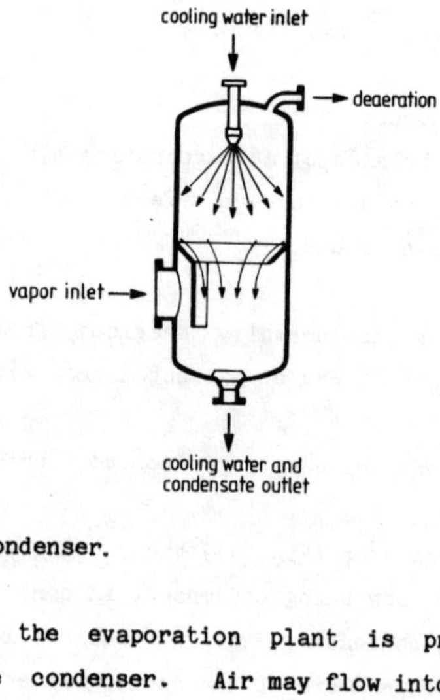

Fig. 12: Mixing condenser.

The vacuum in the evaporation plant is produced by the vapour condensing in the condenser. Air may flow into a plant which is operating under vacuum via leaks at flanges, fittings, covers, manholes etc. and non-condensable gases may enter the plant with the product. Hence, additional measures are necessary to remove the non-condensable gases from the system. Wherever inert gas may accumulate, de-aeration pipes should be provided to draw it off and lead it to the condenser. If not removed from the plant, these gases impair the condensation and interfere with vacuum production.

Liquid ring vacuum pumps and steam jet air ejectors are used as vacuum pumps. Steam jet air ejectors are of simple design and reliable in operation, but require more energy. However their use can be justified if the steam can be used subsequently for preheating the product.

Liquid ring vacuum pumps need less energy for vacuum production than steam jet air ejectors, but they require a cold water supply and need more maintenance.

There are other process steps which influence evaporation, e.g. concentrate cooling, integrated air preheating for dryer air, and CIP cleaning using spray nozzles in the calandria base, the separator channels and the separators. However these have less influence on evaporation than the operations mentioned earlier and no detailed description will be given here.

SUMMARY

In summary, evaporation consists of a number of separate process steps. The designer does not aim at the optimum performance for each individual piece of equipment. Instead the optimum overall design has to be found as a compromise between diametrically-opposed factors.

The computer can be used in the design of individual components and of the overall plant. However, the computer system alone can not find the optimum design solution, and it must be supplemented by acquired experience and human judgement.

APPLICATIONS OF MEMBRANE PROCESSING IN THE DAIRY INDUSTRY

B. OSTERGAARD
Pasilac A/S, DK 8600 Silkeborg, Denmark

ABSTRACT

Since membrane filtration techniques were introduced in the late sixties, considerable advances have been made with membranes, modules, and process technology. High-resistance membranes with different properties have been developed and there have been important advances in the development of high-concentration plate and frame ultrafiltration (UF) modules.

The main applications for UF in the dairy industry are the production of whey protein concentrate (WPC) and cheeses.

Many different cheeses are now being produced by means of UF, and a major development program now underway for production by UF of the ripened types of cheese (yellow cheeses) will bring about a radical change in cheese technology in the next decade.

INTRODUCTION

The introduction of membrane filtration technology about fifteen years ago was a considerable innovation in the dairy industry, and today many industrial UF and Reverse Osmosis (RO) plants are in operation producing various products. This paper will deal mainly with ultrafiltration: firstly with some membranes and modules now available, and secondly with applications of UF in the dairy industry. The main emphasis in the paper will be on production of cheese by UF techniques, comparing it with conventional cheese manufacturing methods, describing manufacturing procedures, discussing various UF cheeses and cheese base, and looking at the probable impact of UF on cheese technology in the future. At the end of the paper there will be a brief discussion on possible ways of using UF permeate, and on dairy applications of RO.

Membranes

The membranes used today are made of polysulphone which is characterised by high resistance to chemicals and temperature and gives excellent results in ultrafiltration of various types of milk and whey.

TABLE 1

Characteristics and Properties of UF Membranes

Membrane type	Approx. cut-off value MW	Recommended operating range			
		pH	Deg.C	bar	MPa
GR51PP	50,000	1-13	0-80	0-10	0-1
GR60PP	25,000	1-13	0-80	0-10	0-1
GR61PP	20,000	1-13	0-80	0-15	0-1.5
GR81PP	6,000	1-13	0-80	0-15	0-1.5

As can be seen in Table 1, the membranes are made with different cut-off values.

They are applied as follows:

GR51PP: UF of acidified skim milk for production of quarg, pate fraiche etc.

GR60PP: UF of acid whey for production of whey protein concentrate. UF of acidified whole milk and cream for production of whole milk quarg, and cream cheese.

GR61PP: UF of sweet whey for production of whey protein concentrate. UF of skim milk/whole milk for production of cultured milk products and cheese.

GR81PP: Normally not used in the dairy industry.

Modules

In concentrating whole milk and skim milk, a very rapid increase in viscosity occurs at about 14% protein for whole milk, corresponding to approximately 34% total solids (TS), and for skim milk at about

18% protein, corresponding to approximately 24% TS. When viscosity is increased, the filtration rate is reduced, thus increasing the risk of blocking the modules. However, milk and whey concentrates are pseudo-plastic, i.e. the apparent viscosity decreases with increased pumping rate.

UF modules have been developed in the light of these viscosity characteristics. The design of the plate ensures a uniform distribution of flow over the complete membrane, flow velocity is high across the membrane, and the tendency to fouling is reduced.

TABLE 2

Maximum total solids in U.F. Plant

Percent TS	A*	B*
Whey, 35% WPC	35	-
Whey, 80% WPC	22.5	35
Skim milk	28	39
Whole milk	40	52
Acidified skim milk	-	22
Acidified whole milk (45% fat/TS)	-	30
Acidified whole milk (60% fat/TS)	-	40
Acidified cream (70% fat/TS)	-	45

* A: Module 36, B: Module 37, DDS/Pasilac.

The development of these modules has made it possible to concentrate to the solids level required for many cheeses, to concentrate acidified milk, and to increase the total solids content in WPC-80 [80% protein in TS] before spray drying. Optimum plant operation is achieved by using module A at lower concentration and module B at higher concentration.

MAIN APPLICATIONS OF UF

The possible applications of UF in the dairy industry are many and diversified:

1. Protein standardization of market milk and milk for powder production.
2. Production of baby food and other nutritional products.

3. Production of whey protein concentrate with protein contents from varying from 35 to 80% protein in TS.
4. Production of cultured milk products such as yoghurt, ymer, and labneh.
5. Production of many different cheese types (see table 3).

Protein Standardization: By adding retentate or permeate as appropriate, protein levels can be standardised to a lower or higher level than the original milk. Fully automated integrated systems for continuous standardization of protein and fat simultaneously are available. Protein standardization makes it possible to produce products with the same protein levels all the year around, and should be as widely used as fat standardization.

Whey Protein Concentrate: Conversion of whey to WPC is the largest single application of UF within the dairy industry, with plants treating from several hundred up to 1 million kg of whey a day, usually to 35%, 60%, or 80% protein/TS. The advantage of UF is that it separates and concentrates without precipitation of whey proteins, and thus conserves their functional properties which have uses in the food industry. The largest field of application - especially for WPC-80 - is in the meat industry, where its excellent water binding and emulsification properties and protein enrichment are utilized.

PRODUCTION OF CHEESE BY ULTRAFILTRATION

Comparison with conventional methods

In conventional cheesemaking, the milk is first coagulated by adding rennet. Concentration takes place when the whey is drained off the curd formed. The whey contains approximately 25% of the protein and 10% of the fat content of the milk, which means that only 75% of the protein and 90% of the fat in the milk are utilized in conventional cheesemaking.

When using UF, the milk is first concentrated by UF, followed by coagulation by means of rennet so that 100% of the fat and 95% of the protein in the milk are utilized. In addition to the large saving in milk consumption (up to 30%), there is also a considerable saving in rennet and starter culture.

Procedure for UF Cheese manufacture

Four different methods are used for producing cheese using UF technology:

1. Protein Standardization to 3.6% to 4.0% protein. This method is more commonly used in France, and it is found that a more homogeneous and uniform cheese quality is achieved all the year round with reduced casein losses.

2. Preconcentration prior to traditional cheesemaking. This technique can be used for a wide variety of cheeses. A typical level of preconcentration is twofold, so that the capacity of the traditional cheesemaking equipment is doubled, and the consumption of starter culture and rennet is halved.

3. Part Concentration to between 20 and 40% total solids (TS). This method is used mostly for cheeses with an open structure. Some changes in traditional cheesemaking equipment are required:-

- Dosing and mixing equipment for retentate, starter culture, and rennet.
- Continuous coagulation equipment.
- Cutting equipment for making cheese grains.
- Other equipment depending on cheese type.

The reason why the milk is only partly concentrated for most cheeses with an open structure is that the cheese grains cannot melt together if the milk is totally concentrated before cheesemaking.

4. Total concentration to a total solids level equal or almost equal to that in the final cheese. This method is the most profitable as the whey proteins are fully utilized. Non-traditional cheesemaking equipment including dosing, mixing, coagulation, and moulding equipment are also required for this method.

Cheeses produced from fully concentrated milk are mainly of the closed structure type.

TYPES OF UF CHEESES

Because of the whey protein content, UF-cheeses naturally differ in composition from traditional cheeses. Nevertheless in regard to appearance, consistency, and taste, UF-cheeses can be:

- Identical or almost identical to the traditional types.
- Different, but still be accepted as traditional types.
- Entirely new, but adjusted to a traditional or non-traditional use.

TABLE 3

UF-cheese types.

Cheese type	Closed Structure	Open Structure	Part-conc.	Full-conc.	UF + evap.
Quarg	X			X	
Pate Fraiche	X			X	
Cream Cheese	X			X	
Cast Feta	X			X	
Domiati	X			X	
Ricotta	X			X	
Cheese Base	X			X	X
Structured Feta		X	X		
Mozzarella		X		X	
Queso Fresco		X	X		
Quartirolo		X	X		
Blue Cheese		X	X		

Cast Feta, Domiati, and Cheese Base are entirely new types of cheese developed on the basis of UF-technology.

Quarg - Fresh Cheese. This group includes skim milk quarg, whole milk quarg, Baker's Cheese, Fromages Frais, Pate Fraiche, and similar types of soft cheeses.

In the initial stages of using UF for these cheeses, a bitter flavour was detected. However, when it became possible to concentrate acidified milk with a pH value of 4.5-4.6, the off-flavour was eliminated most probably because of the higher solubility of calcium at a low pH-values, whereby the calcium leaves with the permeate.

Today, several UF-plants are in operation for production of these products, and many more will be constructed in the next few years.

Using UF technology for skim milk quarg production results in a saving of 10-20% in milk consumption, depending on the traditional method used.

Feta and Domiati. The first cheese produced by UF was Feta. Production began in Denmark in 1976, and today many UF-plants are in operation producing Feta and Domiati. Cast Feta has been very successful although it differs from the traditional Feta which has an open structure.

140,000 tons of Feta and Domiati were produced in Denmark alone in 1984.

Domiati is an Egyptian variant of white cheese. The milk is concentrated by UF in the same way as for Feta. It is then acidified and packed directly into its retail package, usually plastic bag or Tetra Brik, in which coagulation takes place without whey separation.

Mozzarella belongs to the group of "Pasta Filata" cheeses, characterized by a structure which resembles that of chicken breast. Mozzarella was originally an Italian cheese, but is now produced in many countries, particularly the United States. The most common type is the low moisture type with 48-55% TS and 30-40% fat/TS. This is produced by full concentration of skim milk using UF, followed by addition of 80% cream before coagulation and further processing. Mozzarella is mainly used in cooking, especially for pizzas.

Considerable quantities of Mozzarella are being produced using UF.

Queso Fresco and Quartirolo are Spanish/South American types of fresh cheeses, also increasingly produced by UF.

Blue Cheese comprises cheeses like Roquefort, Danablue, Gorgonzola, and Stilton, and from this group Danablue can now be produced by UF with excellent results.

PROCESS TECHNOLOGY FOR CHEESE BASE

Cheese Base (Low Moisture) is an entirely new product developed to partly or fully replace Cheddar, Danbo, Samsoe, Gouda, and other cheeses in the production of Processed Cheese and Cheese Spreads. Its total solids is usually 58-62%, fat/TS 45-50%, and pH 5·2.

Figure 1 shows a process diagram for Low Moisture Cheese Base.

Process: Whole milk is standardized to the required fat in solids, pasteurized at 72°C for 15 seconds, and cooled either to 50°C for UF, or to 2 to 4°C for intermediate storage. The ultrafiltration is carried out continuously at 50°C, until the retentate has reached a total solids content of 40%.

To reduce the lactose content in the retentate, diafiltration can be included in the process until the lactose content is reduced to 2%, at which level the acidification in the final cheese will stop at the desired pH-value of 5.20 +/- 0.05. If diafiltration is excluded, it reduces the capital cost but makes control of the final pH-value more difficult.

Because the bacteria count in the retentate increases during ultrafiltration because of the concentration effect, it is pasteurised at 72°C for 15 seconds. It is then cooled to 30°C and pumped to processing tanks, where starter, NaCl and a small amount of rennet are added. The starter is normally a standard Cheddar culture. Pre-acidification takes place for approximately 2 hours to ensure that sufficient culture survives in the evaporator. The pH-value should be 6.3-6.4 before evaporation.

The processing tanks also serve as balance tanks, to allow for variations in the throughput of the UF-plant and to ensure a constant flow-rate to the evaporator. The heating surface of the evaporator is best utilized when the temperature of the retentate feed to the evaporator is 43°C. The retentate is evaporated under vacuum at a maximum product temperature of 44°C. At maximum solids concentration additional mechanical action is required to overcome high viscosity. This causes a release of free fat.

The product is pumped from the evaporator, to a cylindrical mould, from the outlet of which the Cheese Base is cut into blocks. vacuum packed, and cartoned.

The Cheese Base is stored at approximately 20°C, until a pH of about 5.2 has been reached, after 1-2 days. It is then cooled to 4°C at which temperature it has a shelf-life of several months.

Cheese Base (High Moisture). This Cheese Base is produced with a total solids content of 47% and 45% fat/TS.

FIGURE 1

Process diagram for Cheese Base

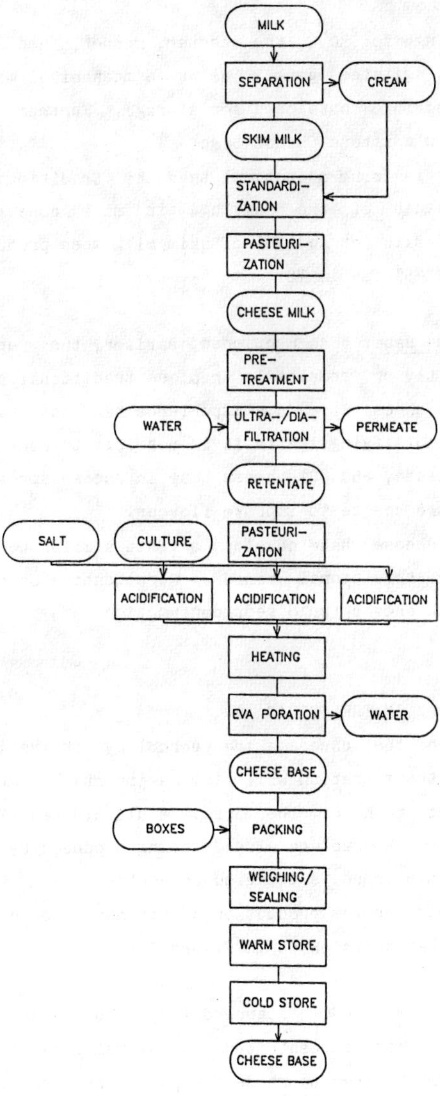

It was developed to simplify the process and to eliminate the capital cost of the swept surface evaporator. Otherwise, the process is the same as for low moisture cheese base, except that the milk is concentrated to 47% TS in the UF-plant. Diafiltration is included in the process to reduce the lactose content of the retentate to approximately 2%.

Following cooling to 30°C, the starter, rennet, and NaCl are added, and the mix is agitated gently using a mechanical mixer. The product is then pumped into containers for storage. Further treatment is the same as for low moisture Cheese Base.

The UF-process is more economical than the traditional cheese process, mainly because of the saving in milk consumption of approximately 1.7% of milk fat and 15% of skim milk when producing a product of 57% TS and 45% fat in TS.

The Use of Cheese Base. As mentioned earlier, the purpose of cheese base is to partly or completely replace traditional cheeses in processed cheese and cheese spreads. Experience has shown that with the right choice of emulsifying salts it is possible to use 50% cheese base in processed cheese, and 90% cheese base in cheese spreads. The remainder can be mature cheese to improve flavour.

When matured, cheese base develops a taste similar to that of cheese produced in the traditional manner. Hence cheese base may soon be developed into new cheese for direct consumption.

CHEESEMAKING TECHNOLOGY IN THE 1990'S

The development of the cheesemaking technology for the 1990's has already begun, and ultrafiltration will play an important part. It has already been seen that much of the Soft, Mould Ripened, Fresh, and Pasta filata types of cheeses are already being produced by means of ultrafiltration and this trend is expected to continue.

A process diagram for the production of Ripened Cheeses (Yellow Cheeses) by ultrafiltration is shown in Figure 2.

Pre-cheese: Pasteurized, standardized, and possibly pre-acidified cheese milk is concentrated by ultrafiltration and by diafiltration if it is required to raise the pH of the cheese.

FIGURE 2
Process diagram for Ripened Cheeses

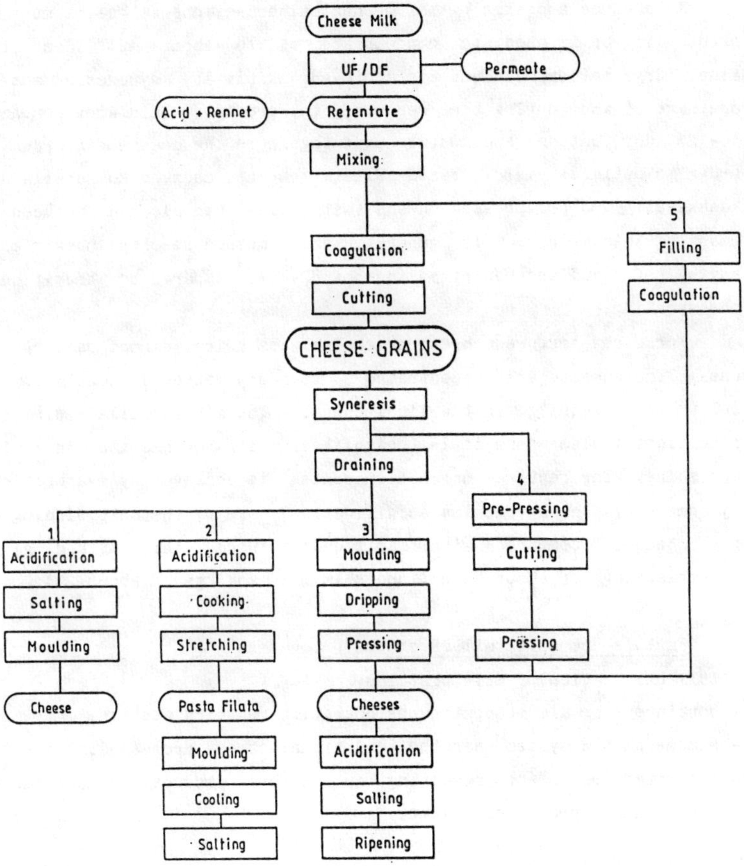

1. Barrel Cheese, Cheddar
2. Mozzarella, Kashkaval etc.
3. Havarti, Esrom, Montery Jack, Feta, Danablu etc.
4. Danbo, Gouda etc.
5. Danbo, Saint Paulin, Camembert etc.
   Feta, Domiati, Akawi

After concentration, the retentate is pasteurized and cooled to the coagulation temperature. Starter, mould culture, possibly other secondary materials and rennet are added and thoroughly mixed. After coagulation, the product is called Pre-cheese.

Process: The mix can now be treated in two ways:-

To produce a cheese with mechanical holes, cheese grains must be produced. UF is used to concentrate up to about 40% TS and the final dry matter content is achieved mainly by syneresis and/or drainage of whey during the rest of the process. This whey contains 10-12% dry matter and must be used again in the cheese in order to order to optimize yield. Pasta Filata cheese, such as Mozzarella and Kashkaval, and pre-pressed cheese (with holes) can also be produced by means of this method. In general, this method results in skim milk saving of 7-10% and rennet savings of 80-85%, compared to traditional cheesemaking.

For close-textured cheese or cheese with holes, direct casting is used. The cheeses are concentrated up to a dry matter content 5 to 10% below that required in the final cheese. The mix is filled directly into tight moulds where it is coagulated. It now has its final size and shape. The required dry matter content is achieved by evaporation, by some whey drainage during acidification, and by the salt binding of the cheese. Compared with traditional methods, this method gives rennet savings of about 80-85% and skim milk savings of about 20%.

Factors common to all UF cheese processes:

- full utilization of milk proteins,
- continuous or almost continuous operation reducing wastage,
- a more closed system than the traditional cheese processes,
- automated equipment resulting in uniform product, reduced labour input, and improved efficiency.

REVERSE OSMOSIS

In UF (working pressure 6-8 bar), the product is both concentrated and separated. With RO however, only concentration occurs, as the membranes allow only almost pure water through. Working pressure is 25-40 bar. While RO may replace evaporation for certain applications, in general it must be regarded as a supplement to other concentration techniques. Its main fields of application in dairying are:

1) Concentration of milk or whey to save transport costs.
2) Pre-concentration in powder production.
3) Concentration of UF permeate in lactose and alcohol production.
4) Concentration of skim milk for ice cream production.
5) Concentration of rinse water, i.e. the first product rinse water from tanks, pipes, and other equipment.

UTILIZATION OF UF PERMEATE

Permeate from RO is very clean and is therefore often used as process water or for cleaning. UF permeate on the other hand is highly polluting with a BOD-value of 25,000-40,000 mg oxygen per litre, and therefore cannot be disposed of as waste water. It contains approximately 5.7% TS of which some 85% is lactose. It is used mainly for animal feeding purposes, for permeate powder, and for various lactose products. New areas of application are fermentation to alcohol, biogas production and hydrolysis to glucose and galactose for use as sweetener. Its use in the production of polyurethane foam has also been recently reported. Further research is needed in this area.

CONCLUSION

UF opens up new prospects of better utilization of milk. It provides for considerable reductions in milk consumption for the production of existing products, and opens the way for the manufacture of new products tailored to the needs of other branches of the food processing industry.

In the production of cheese using UF techniques, the financial savings are substantial, and typical pay-back times vary between five and thirty-six months.

However increased productivity and economy are not the only relevant factors in deciding whether to adopt new technology in the dairy industry - the maintenance or improvement of hygiene standards and product quality are also essential and these are guaranteed by ultrafiltration technology.

ENERGY ASPECTS OF FOOD PRECONCENTRATION

H. G. Kessler
Institut für Milchwissenschaft und Lebensmittelverfahrenstechnik,
Technische Universität München, D-8050 Freising-Weihenstephan

INTRODUCTION

By preconcentration is meant the removal of water from liquid foods and not the increase in concentration of certain food constituents such as that of proteins by ultrafiltration or that of butterfat by centrifugation.

Fig. 1 depicts three methods that can be used for the removal of water: the removal as a solid in the form of ice by freeze concentration, the removal as a liquid in the form of water by reverse osmosis and as a gas in the form of vapour by evaporation. Schwartzberg (ref. 1) has discussed the various methods used for food concentration and their energy requirements.

Much development work has been done on these methods in recent years with special emphasis on their energy requirements without neglecting, of course, product quality. In this paper, the current state of the technology will be described, along with a discussion of a number of problems one meets in processing, such as the limitations of reverse osmosis and the formation of product deposits in falling film evaporators.

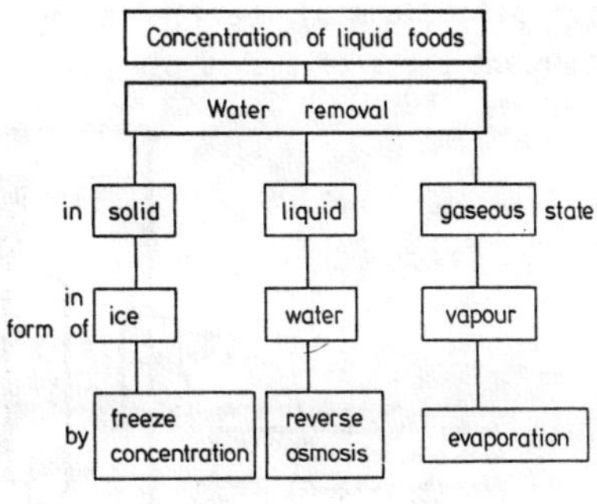

Fig.1 : Methods for concentration of liquid foods

In order to make the energy requirements of the various processes comparable, the electrical and heat energy consumed in each process was converted to primary energy and calculated as the energy required for the removal of 1 kg of water. For that purpose the electrical energy consumed was multiplied by a factor of 3 and the heat energy by a factor of 1.2.

Freeze concentration

Freeze concentration is most suitable for fruit juices or extracts with highly volatile aroma constituents. When water-containing foods are cooled below their freezing point, pure ice crystals are formed. If the ice crystals are separated, a product with a higher total solids content is obtained. The freezing point, i.e. the temperature at which the first ice crystal formation takes place, is lower the higher the concentration of dissolved solids in the product. Fig. 2 shows this for some products (ref. 2).

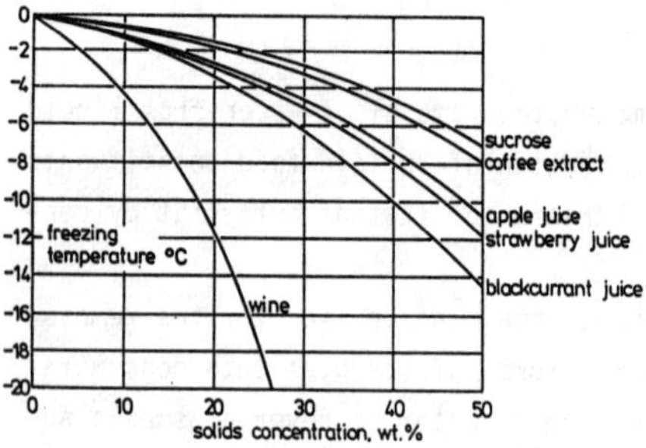

Fig.2: Freezing point curves (Thijssen, 1969)

The diagram of a freeze concentration plant made by the firm Grenco (ref. 3) in Fig. 3 depicts the most important parts of such a plant.

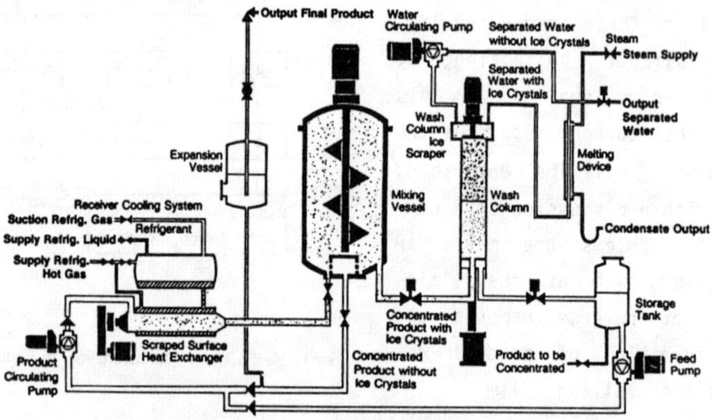

Fig. 3: Freeze concentration plant (Grenco)

These are: a scraped surface heat exchanger in which the heat is extracted for freezing, a mixing vessel in which the ice crystals are allowed to grow and a wash column in which the ice is separated from the concentrate by filtering and washing with ice water. There has been much success in recent years with the growing of large ice crystals. This has so improved the separation of ice and concentrate that hardly any product is now lost. With the older methods, solids concentrations of no more than 35 % - 38 % were economically achievable, but concentrations of 40 % to 45 % can be obtained nowadays.

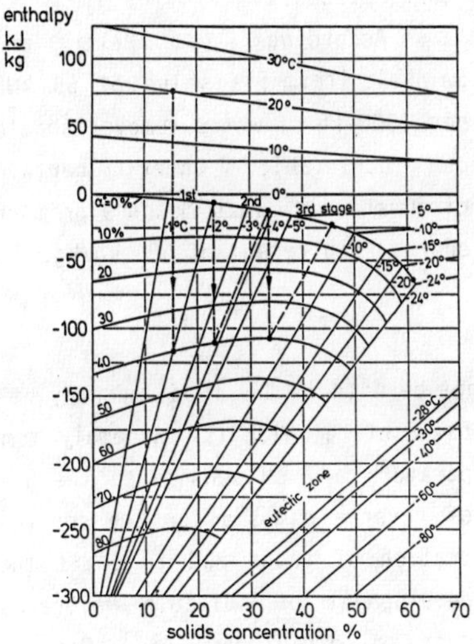

α* = proportion of ice based on the amount of water

Fig. 4: Enthalpy-solids concentration diagram for coffee extract

The effect of the solids concentration on precipitation by freezing characterized by the ice-curve, enthalpy, temperature and proportion of ice, is shown in Fig. 4, for coffee extract. This graph makes it possible to read off directly enthalpy differences, the proportion of ice formed and the solids concentration of the concentrate when the temperatures are lowered. The diagram shows an example of a three stage concentration. Starting with an initial solids concentration of 15 % the extract is cooled to -2°C. Then 40 % of the total water forms ice crystals which are separated from the concentrate at 24 % solids concentration.

The second stage consists of cooling to -3.8°C when 40 % of the water present at the beginning of stage 2 forms ice crystals. After the separation of these the solids concentration has been raised to 34 %. The third stage produces a final concentration of 46 % at a temperature of -7.5°C. The amount of heat extracted in the process can be read off directly from Fig. 4. It is 195 kJ/kg in the first stage, 100 kJ/kg in the second stage and 90 kJ/kg in the third stage.

The energy required for freeze concentration in single stage and 4 stage plants has been compared by Pelt (ref. 4) (Fig. 5)

All plants use the heat of fusion of the ice crystals to condense a part of the hot gases in the compressor. These studies have shown that the use of multistage plants with appropiately high capacities make it possible to reduce energy requirements considerably. However, with an

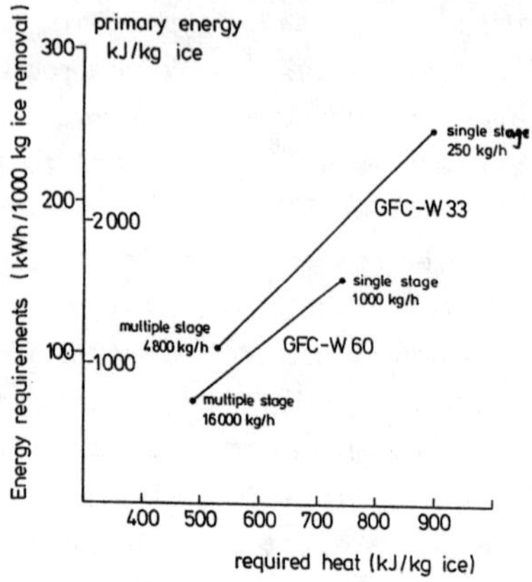

Fig.5: Total energy requirements per 1000 kg water removal in freeze concentration (van Pelt, Grenco,1983); $T_{Condenser} = 30°C$

energy consumption of 69 kWh per 1000 kg of water removed - which corresponds to 248 kJ/kg or three times as much as the primary energy, namely 745 kJ/kg - the process compares unfavourably with other methods.

According to Pelt and Swinkels (**ref.** 5) a value of 55 kWh per 1000 kg of water removed should be achievable under certain circumstances. This means a primary energy requirement of 594 kJ/kg.

Reverse osmosis

For a long time reverse osmosis could not be used in the food industry because the membranes were made of cellulose acetate which is **extremely** temperature and pH sensitive. The use of reverse osmosis for the concentration of foods had to await the development of suitable membranes such as polysulphone ones. Reverse osmosis plants can now be cleaned by the usual plant cleaning methods.

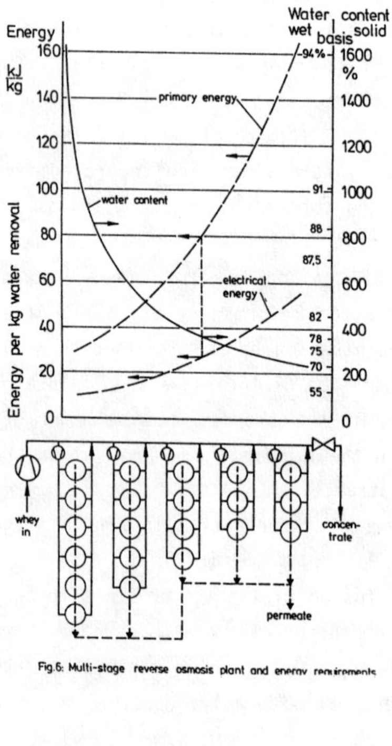

Fig.6: Multi-stage reverse osmosis plant and energy requirements

The reverse osmosis modules on the market nowadays are very similar to the ultrafiltration ones, i.e. tubular, plate and spiral modules. In contrast to ultrafiltration, reverse osmosis does not involve a filtration in which the permeate passing through the membrane contains salts, lactose etc. In RO the permeate passes through the membrane by solution diffusion. Because the aqueous liquid to be concentrated is at a pressure which is greater than the osmotic one, only pure water passes, or rather should pass as permeate through the membrane. How far this ideal state is achievable will be discussed later.

Fig. 6 is an schematic representation of a multistage arrangement for the concentration of, for instance, whey.

Energy is only required for the operation of the pumps. There is a high pressure pump which feeds the product to be concentrated into the system and maintains the required pressure by means of a pressure regulating valve, and there are also circulating pumps. Depending on the degree of concentration required, the plant may be operated as a multistage one, i.e. several circulation systems arranged in series. Each system is equipped with RO modules arranged in parallel and with a product circulating pump.

The abscissa in Fig. 6 represent the serially arranged stages of a RO plant. Energy requirements and the water content of the concentrate are plotted as the ordinate. The energy figures were obtained from the few experimental results and literature reports available at present. The water content figures were related to total solids content because this is a measure of the amount of water removed.

Fig. 6 shows that a relatively large amount of water is removed in the first stage with a relatively low energy consumption per kg of water removed. The flux, i.e. the specific permeation rate, decreases in the following stages, the degree of concentration does not increase very much any more and the energy consumption becomes correspondingly very large. In the example illustrated in Fig. 6 about 80 kJ of primary energy per kg of water removed were needed to concentrate whey from 94 % moisture to 78 % moisture content (wet basis).

Although in the example almost 80 % of the total initial amount of water has been removed by RO, the question arises what degree of concentration can reasonably be achieved by RO. Are the high concentrations associated with limiting factors, suchs as too low flux or high viscosities, and/or is too much dissolved solids passing through the membrane.

Some measurements have been carried out at our Institute using a small tubular plant (Patterson Candy International Ltd.). Fig. 7 shows the dependence of the flux on the solids concentration of the retentate during the concentration of skim milk at constant pressure and temperature. The only variable was the rate of flow in the tubes (d = 12.5 mm Ø).

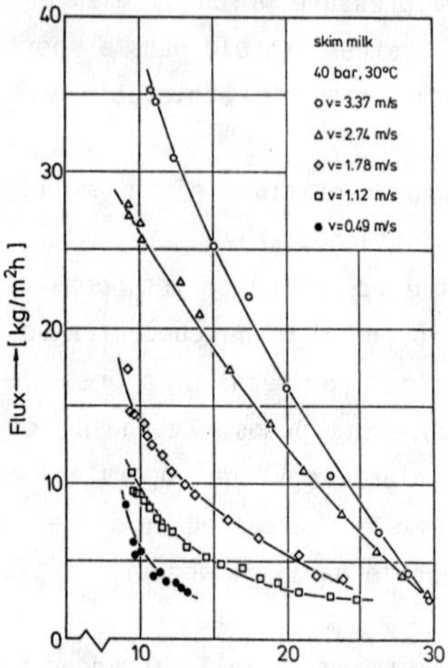

Fig.7: Dependence of the flux on solids concentration

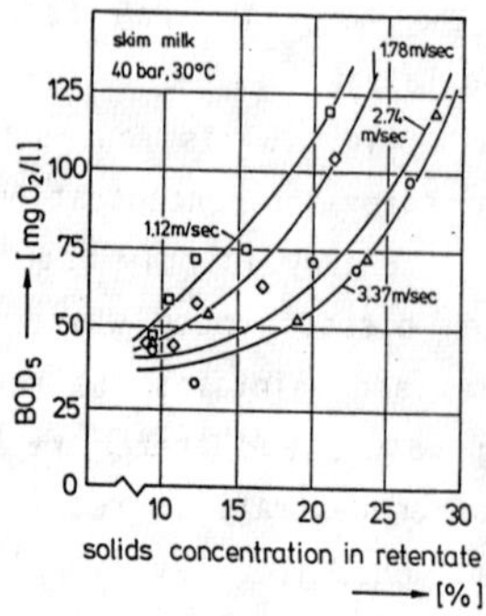

Fig.8: Dependence of BOD_5 of the permeate on solids concentration

It is seen that high flow rates markedly increased the flux because high shearing forces reduced the thickness of the deposited layer. However, when the degree of concentration became too high the flux became so low so as to make the operation uneconomical. It may be concluded from these results that it would be desirable to produce high shearing forces with as little energy expenditure as possible.

In these experiments the relationship between the solids concentration of the retentate and certain properties of the permeate were examined. Measurements of the BOD_5, the electrical conductivity and the freezing point (depicted in Figs. 8, 9, and 10) indicated that the amount of dissolved substances passing through the membrane and being lost in the permeate increased as the solids content of the retentate increased. This is important not so much because of the loss of product but because the load on the effluent plant might become excessive.

The measures necessary to counteract this can be read off the graphs. Namely, the concentration of the dissolved solids in the layer adjacent to the membrane should be as low as possible to keep the concentration gradient, which is responsible for diffusion from the inside to the outside, small. This is the case when the solids content of the retentate is low. But an increase in the shear stress produced by higher flow rates also lowers the concentration of solids in the layer next to the membrane. This manifests itself as a decrease in the amount of solids passing through the membrane at constant solids concentration in the bulk flow.

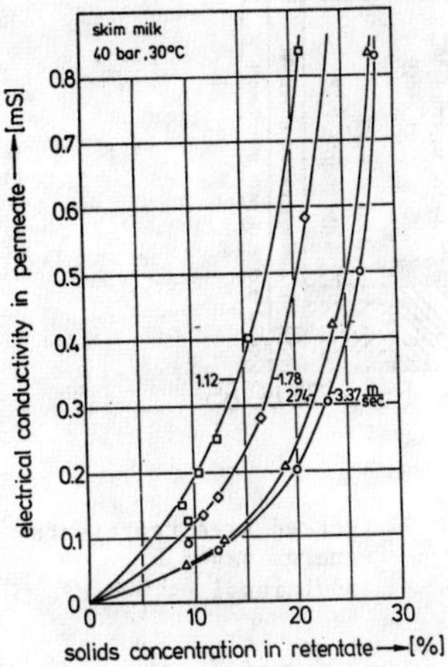

Fig.9: Dependence of electrical conductivity of the permeate on solids concentration

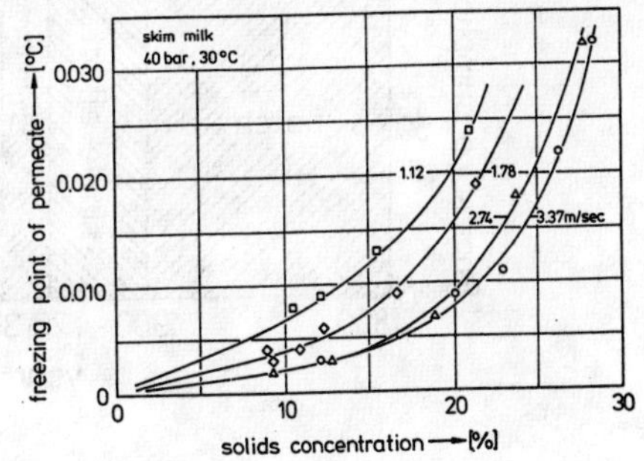

Fig.10: Dependence of the freezing point of the permeate on solids concentration

Because the dissolved solids penetrate the membrane, driven by the difference in concentration, but water permeates because the applied pressure is greater than the osmotic pressure, a further increase in pressure will increase the flux of water and reduce the proportion of dissolved solids in the permeate.

Evaporation

The demand for ever higher quality concentrates and milk powders as well as the increase in the volume of production of those products has led to the development of new evaporation equipment. The batch evaporator operating under vacuum was replaced by the circulation evaporator and that in turn was succeeded in about 1955 by the falling film evaporator which is nowadays practically the only equipment used for concentration in the dairy industry (Fig. 11).

Energy requirements became a serious problem after the two energy crises of 1972/1973 and 1978. Although the technique of multiple effect evaporation had been known in the sugar industry before 1900, it

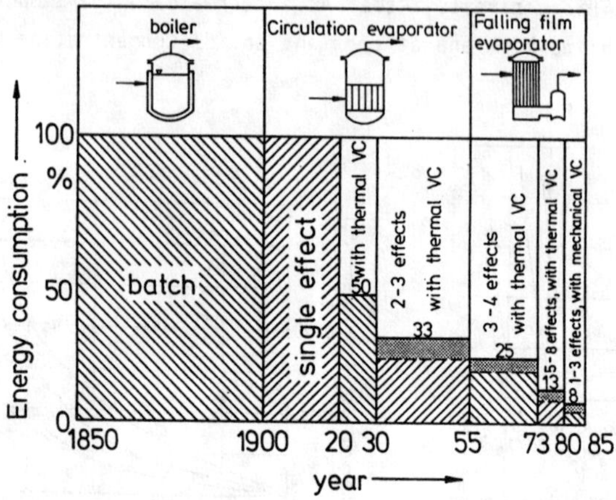

Fig.11: Development of evaporation plants

Dotted area represents energy saved by additional effect(s)

could not be used effectively in the dairy industry before the introduction of the falling film evaporator which operates with small temperature differences. Another invention for the saving of energy, the thermal vapour compressor had also been known as early as about 1900 and plants with mechanical vapour compression had been operating in Switzerland for several years before 1945. However, because energy was readily available after 1950, this capital intensive technology fell into disuse. The energy crises mentioned above brought about, first of all, an increase in the number of effects in evaporators,culminating in the development of plants with eight effects. In 1979 mechanical vapour compression was used once again in the Federal Republic of Germany in a triple effect plant for the evaporation of whey. The most recent developments are the change from the expensive radial compressors to simple fans. Because pressure or temperature differences of no more than 4 - 5 K can be achieved with these machines, only single effect plants with subdivided evaporators are practicable unless a separate compressor is provided for each effect.

Some schematic representations of equipment for the evaporation of milk and whey illustrate the development since 1955.

Fig. 12 is a diagram of a triple effect evaporator with thermocompression such as was used in the whole of the dairy industry until 10 years ago. At temperature differences of 15°C the compressor can

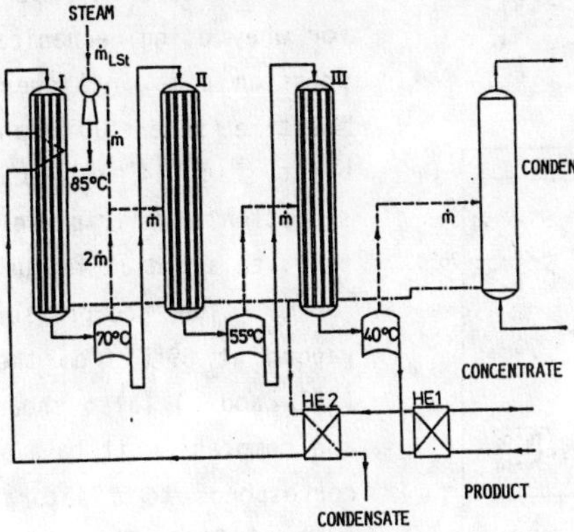

Fig.12: Triple effect evaporator with thermorecompression
Spec. electrical energy: 18 kJ/kg water removal
Spec. steam energy: 736 kJ/kg water removal
Spec. primary energy: 1000 kJ/kg water removal

suck in about the same amount of vapour with a given amount of live steam so that the steam consumption of this arrangement is 25 %. In practice this ratio is higher, namely approximately 0.32, since more steam is needed to preheat the product and to drive the steam jet vacuum pumps. The cleaning process and the starting-up of the plant must also be taken into account.

The enthalpy/entropy diagram of Fig. 13 illustrates the operation of the steam jet vapour compressor and the amount of vapour that can be sucked in by it.

The ratio of the mass of vapour to that of the live steam can be calculated from the differences in enthalpy arising from the expansion of live steam and the compression. As this ratio is 1 when the temperature of the saturated steam is raised by 15°C, a smaller temperature difference of e.g. 3°C will make it possible to suck in and compress more than 3 kg of vapour with 1 kg of live steam.

Measures such as the reduction in temperature differences (3 to 4°C) and an increase in the number of effects (up to 8) increase the cost of the plant considerably but make it possible to reduce the energy consumption to about 10 % of the steam used, which for the whole plant corresponds to about 400 kJ of primary energy per kg of water removed.

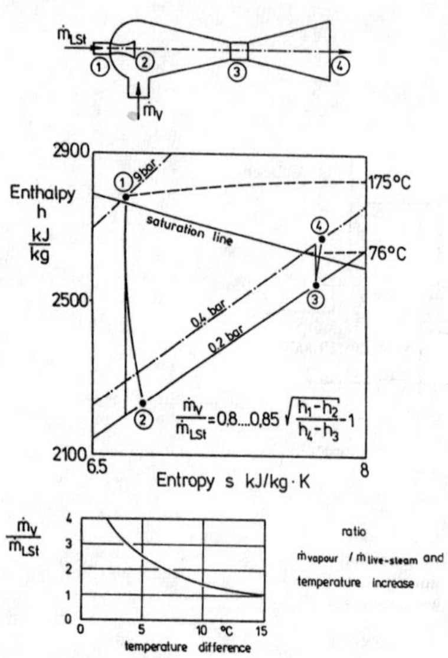

Fig.13: Steam-jet vapour recompressor and h-s-diagramm

In 1979 the first evaporating plant for whey using mechanical vapour compression came into operation. In the meantime some 100 plants have been built. Fig. 14 is a schematic representation of a triple effect evaporator with a radial vapour compressor.

The compressor sucks in vapour at 59°C from the last effect (IIIa and b) (also shown in Fig. 15) and compresses it to a pressure which corresponds to a saturation temperature of 71°C. The resulting increase in temperature to 125°C is reduced to 71°C by the injection of condensate. This heat as well as the heat of the condensate and of the concentrate are used for the preheating of the product. The introduction of radial compressors made it possible to lower the primary energy consumption to about 200 kJ/kg of water removed. Fig. 15 is an enthalpy/entropy diagram for the operation described in the **example.**

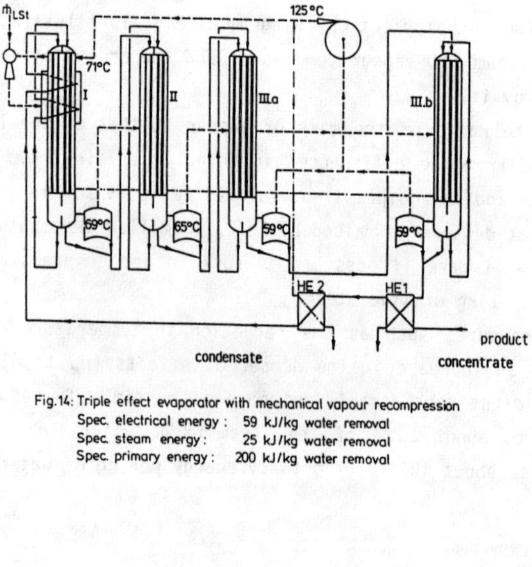

Fig.14: Triple effect evaporator with mechanical vapour recompression
Spec. electrical energy : 59 kJ/kg water removal
Spec. steam energy : 25 kJ/kg water removal
Spec. primary energy : 200 kJ/kg water removal

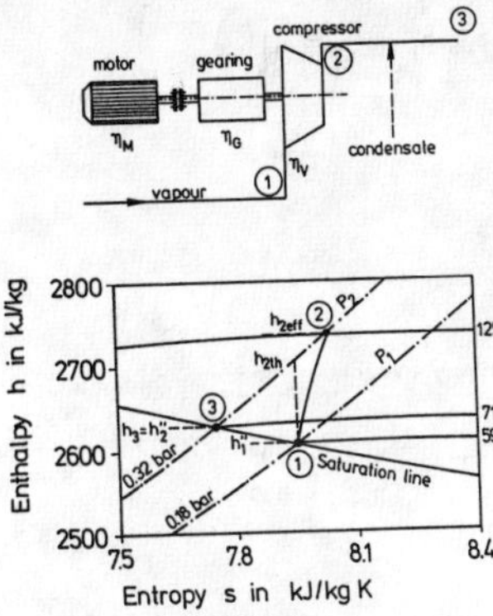

Power of the E-motor

$$P_{el} = \frac{P_V}{\eta_V \cdot \eta_G \cdot \eta_M} = \frac{\dot{m}_V (h_{2th} - h_1)}{\eta_{ges}}$$

Fig.15: Radial vapour recompressor and h-s-diagram

Single effect evaporators in which the vapour is compressed by simple fans (Fig. 16) have been in operation since 1983.

This has made it possible to increase the saturation temperature by 4 - 5°C. The energy consumption can again be obtained from the enthalpy/entropy diagram of Fig. 17.

For plants with smaller evaporation capacities of up to 5 t/h the isentropic efficiency $\eta_{isentrop}$ is about 70 to 75 % and for plants with capacities of up to 30 t/h the isentropic efficiency is about 85 %. In the example in Fig. 17 a value of 80 % was chosen. η_M at 93 % includes frictional losses at shafts and clutches, losses due to variable drives, **frequency** regulators, control panels etc.

When the calculation of the energy consumption for a temperature increase from 60 to 64°C takes into account the energy consumed by

Fig.16: Fan for evaporating system (Anton Piller, Osterode, F.R.G.)

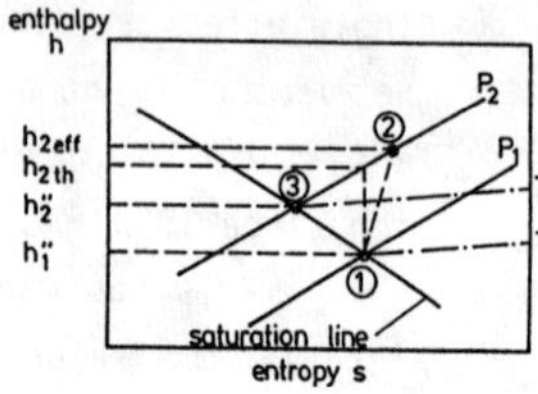

$$h_{2th} - h_1'' = \frac{n}{n-1}\, P_1 v_1 \left[\left(\frac{P_2}{P_1}\right)^{\frac{n-1}{n}} - 1\right]$$

$n = 1.3$

with $T_1 = 60°C$

Temp. diff. °C	$h_{2th} - h_1''$ kJ/kg	primary energy kJ/kg
4	28.5	115
5	35.7	144

$$\text{primary energy} = 3 \cdot \frac{h_{2th} - h_1''}{\eta_M \cdot \eta_{isentr}}$$

$\eta_M = 0.93$

$\eta_{isentr} = 0.80$

Fig. 17: Mechanical recompression by fan

every part of the plant (circulating pumps, vacuum pumps, steam consumption on starting up) then the energy required is composed of the following: Fan drive ~ 38 kJ/kg, other electrical drives ~ 18 kJ/kg, steam heat ~ 25 kJ/kg. This adds up to a total primary energy requirement of ~ 198 kJ/kg of water removed. In principle all mechanical compressors could be driven by gas or fuel motors whose exhaust heat could then be used for heating. This would reduce the primary energy requirement considerably.

As the energy utilization is increased by these innovations, the area required for the plant also increases. As a result of the developments in evaporation technology the well known phenomenon of deposit formation has become a problem. The manufacturers of a modern plant with mechanical vapour recompression and large evaporating surfaces would like to choose high temperatures in order to keep the compressor as

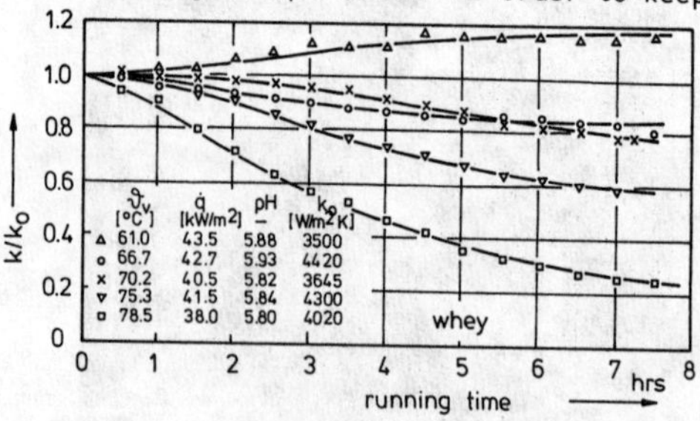

Fig. 18: Relative overall heat transfer coefficient k/k_0 during the evaporation of whey at different evaporation temperatures ϑ_V.

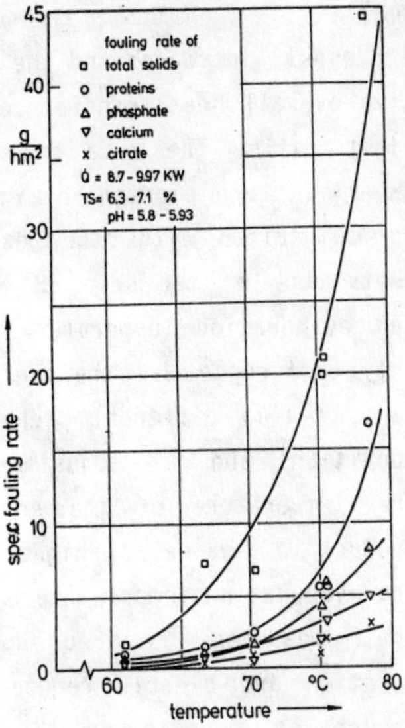

Fig. 19 b: Fouling rates during evaporation of whey — dependence on evaporation temperature

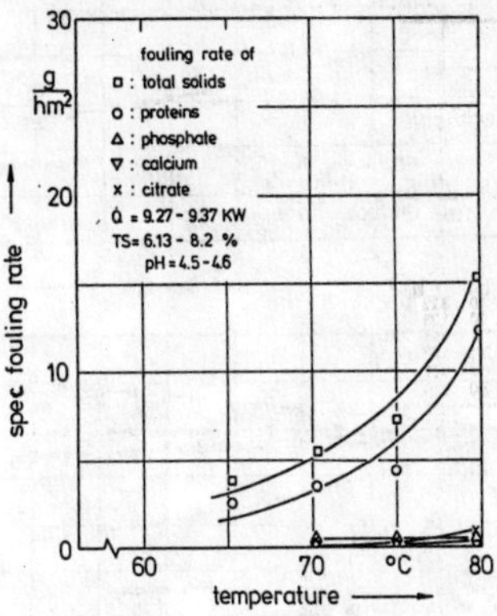

Fig. 19 a: Fouling rates during evaporation of whey — dependence on evaporation temperature

small as possible. But high temperatures combined with high residence times mean increased deposit formation and increased energy consumption.

Figs. 18, 19 and 20 present some results from a more comprehensive investigation at our Institute of the deposit formation in an evaporator tube.

The decrease in the relative overall heat transfer coefficient in whey evaporation is shown in Fig. 18. It is evident that the overall heat transfer decreases as the temperature of evaporation increases. The improvement at 61°C is due to the higher rates of bubble evaporation. This may be explained by an increase in the number of nuclei on the surface of the tube which initiates and promotes the production of vapour bubbles.

Figs. 19a and 19b also show that deposit formation falls with decreasing evaporating temperatures.

If long running times are required then evaporating temperatures should range from about 58 to 60 °C. Figs. 19a and 19b differ in the composition of the deposit. At lower pH values the proportion of salts in the deposit is considerably smaller than at the higher pH value.

Fig. 20a shows that during the evaporation of UF-permeate, i.e. in the absence of proteins, there is no deposit of salts if a low pH value has made the solution unsaturated. When the solution is supersatura-

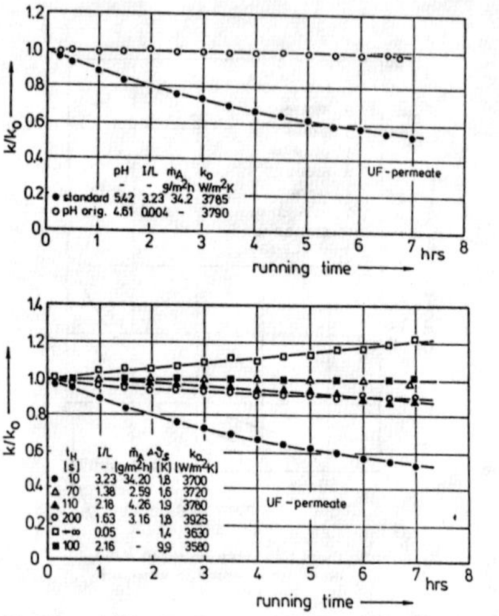

Fig. 20a and 20b: Relative overall heat transfer coefficient k/k_0 during the evaporation of UF-permeate

ted, e.g. at pH 5.42, the amount of deposit increases and the relative overall heat transfer coefficient falls. The data from this experiment are plotted in Fig. 20b for comparison with other experiments done at the same pH values and evaporation temperatures. The only difference was the pretreatment, namely different preheating conditions and the adjustment of the temperature of the entering product to a value $\Delta\vartheta_S$ higher than the evaporating temperature ϑ_V. These measures abolished supersaturation and clearly reduced the amount of deposit or, in some cases, eliminated it completely. The increase in k/k_0 may again be explained by the increase in the number of nuclei on the surface of the evaporator tube and the resulting increase in the rate of bubble evaporation.

SUMMARY

The development of industrial concentration plants for liquid foods was determined by considerations of keeping quality (e.g. condensed milk, tomato concentrate), product quality (e.g. flavoured juices and extracts) and energy savings. The last point is of special interest since 1973, when energy prices increased.

Nowadays three main processes are used to concentrate liquid foods: 1st freeze concentration where water is removed in the frozen state as ice, 2nd reverse osmosis where water is removed in the liquid state as water, and 3rd evaporation with water removal in the gaseous state as vapour.

Freeze concentration is suitable for fruit juices and extracts when concentrates of high quality with less aroma and flavour losses are required. Although there have been improvements in energy saving methods of processing, the specific heat consumption - calculated on the basis of primary energy - is high compared with the other concentration processes.

Reverse osmosis became important during the last 5 years as new membranes, resistant to the temperatures and pH-values used in food processing were developed. Even if it is not possible to obtain concentration degrees comparable to those in evaporators, the process is very economical. But the energy consumption will increa se considerably when higher concentrations are reached. Furthermore, it must be borne in mind that under certain conditions, dissolved constituents penetrate the RO-membrane together with the permeate, thus increasing effluent load.

A great development took place in evaporation plants with regard to energy saving measures. This proceeded via triple effect evaporators to 5 to 7 effect ones with thermal recompression, to triple effect evaporators with mechanical recompression. As a consequence of decreasing temperature differences for both types of recompression, heat exchange surfaces increased. At the moment the most usual evaporation plant is a single effect evaporator equipped with a common fan which makes it possible to operate with a temperature difference of about 4°C. By means of this, investment-, maintenance and energy costs can be kept at a low level. On the other hand, low temperature differences require longer residence times of the product in a plant. But this can increase deposit formation and shorten running times. Therefore investigations were undertaken into the possibilities of reducing deposits, increasing the running times and the efficiency of evaporation plants.

The table summarizes the energy consumption per 1 kg of water removed in the various processes. For ease of comparison the results are

concentration process	spec. electrical energy kJ/kg	spec. steam energy kJ/kg	spec. primary energy kJ/kg
freeze concentr.	55 ... 75	-	~ 600 ... 800
triple effect + therm. recompr.	18	736	~ 1000
multiple effect + therm. recompr.	30	245	~ 400
triple effect + mech. recompr.	59	25	~ 200
single effect + fan	56	25	~ 200
reverse osmosis	15 ... 30	-	~ 45 ... 90

concentration processes and energy consumption in kJ per kg water removal

reported as primary energy i.e. electrical energy was multiplied by a factor of 3 and thermal energy by a factor of 1.2.

From the point of view of energy consumption, freeze concentration should be reserved for special applications. RO requires the least amount of energy, provided that the degree of concentration is not too high. If high degrees of concentration are required, a two-stage process is recommended, with RO as the first stage and a single effect evaporator with mechanical vapour compression by fans as the second one. Thus, with whey, it would be economical to remove 2/3 of the water by RO and 1/3 by evaporation.

NOMENCLATURE

h''_1	(kJ/kg)	Enthalpy of the saturation state 1
h''_2	(kJ/kg)	Enthalpy of the saturation state 2
$h_{2\ th}$	(kJ/kg)	Theoretical enthalpy after compression
$h_{2\ eff}$	(kJ/kg)	Effective enthalpy after compression
h_3	(kJ/kg)	Enthalpy of the state 3
I/L	(-)	Ratio of the ionic to the solubility product
k_0	(W/m^2K)	Overall heat transfer coefficient of the clean evaporation surface
k	(W/m^2K)	Overall heat transfer coefficient of the evaporation surface with deposits
$\dot{m}$	(kg/s)	Mass flow of vapour
$\dot{m}_{L\ St}$	(kg/s)	Mass flow of live-steam
$\dot{m}_V$	(kg/s)	Mass flow of vapour
$\dot{m}_A$	(g/m^2h)	Specific fouling rate
P_1	(bar)	Vapour pressure before compression
P_2	(bar)	Vapour pressure after compression
P_{el}	(kW)	Electrical power
P_V	(kW)	Theoretical power to compress $\dot{m}_V$
$\dot{q}$	(kW/m^2)	Specific heat flow rate
$\dot{Q}$	(kW)	Heat flow
t_H	(s)	Holding time
η_M	(-)	Motor efficiency
η_g	(-)	Gearing efficiency
η_V	(-)	Compressor efficiency
$\eta_{isentr.}$	(-)	Isentropic efficiency
ϑ_{Ges}	(-)	Total efficiency
ϑ_V	(°C)	Evaporation temperature
$\Delta\vartheta_S$	(K)	Difference of temperature

REFERENCES

1. H.G. Schwartzberg, Food Technology, 31 , (1977), 67-76.

2. H.A.C. Thijssen, Dechema Monogr. 63 (1969), 153-177.

3. J.N. Wagner, Food Engineering, February, 1983

4. W.H.J.M. van Pelt, in Grenco Special Applications, B.V., Holland, May 1983.

5. W.H.J.M. van Pelt and W.J. Swinkels, Fourth International Congress Engineering and Food, Edmonton, Alberta, July 7-10, 1985.

HEAT AND MASS TRANSFER ASPECTS AND MODELING OF DRYERS - A CRITICAL EVALUATION

F.W. Bakker-Arkema, Professor
Agricultural Engineering Department
Michigan State University
East Lansing, MI, U.S.A.

ABSTRACT

Drying is still considered an art by many food engineers. This paper shows that the time has come for simulation modeling to make a major contribution to dryer design.

The state-of-the-art of simulation modeling of three dryer types used commonly in the food industry is critically reviewed with respect to its application for commercial dryer design.

The concurrent-flow packed-bed (CCFPB) dryer is modeled by five differential heat and moisture transfer equations. The model has been used successfully in industry for the design of various types of multi-stage CCFPB dryers for commercial rice milling plants.

The concurrent-flow cascading rotary (CCFCR) dryer simulator requires a heat and mass transfer model (similar to that of the CCFPB dryer) in addition to a particle transport model. Although the precise knowledge of some aspects of food particle movement in a CCFCR dryer is still lacking, several particle transport equations available in the literature are considered acceptable for useage in the rotary dryer modeling of such solid particle products as parboiled rice.

The concurrent-flow spray (CCFS) dryer model requires individual models for atomization, particle trajectory transport, airflow distribution and heat and moisture transfer. A CCFS dryer model should be able to predict the two-dimensional nature of two-way coupling of the spray drying process. Although significant progress has been made in the past few years in the modeling of CCFS dryers, additional research is required before modeling can be used for commercial design in the food industry.

1. INTRODUCTION

Drying (dehydration) is an important unit operation in the food industry. Examples are the solid bed drying of parboiled rice, the rotary drying of raw sugar, and the spraydrying of skim milk. Dryer design is considered by many engineers still largely an art (van Brakel, 1978; Purcell, 1979). Although the theory of some phases of the drying process are well understood, other aspects have thus far escaped theoretical interpretation (Reay, 1979). Scale-up problems have further contributed to the lack of application of simulation in the design of dryers for the food industry (Breslin, 1985).

The dehydration of food products requires considerable energy, between 1000 and 2000 kJ per kg of water (Senhaji and Bimbenet, 1984). Operating costs vary from 4 cents to 15 U.S. cents per kg of water evaporated for the most common food dryers (Reshaw et al., 1982). Optimal design of food dryers can result in considerable savings in energy and operating costs. Optimum design by trial and error is time-consuming and costly,but can be accomplished quickly and inexpensively with the use of simulation modeling (Brook and Bakker-Arkema, 1980). The optimization can only be successful if sufficiently accurate dryer models are available.

In this paper the simulation models of three food dryer types are critically reviewed with respect to their availability, suitability, and accessibility as tools for commercial design. Each of the dryers is of the concurrent-flow type: (1) the packed-bed dryer, (2) the cascading rotary dryer, and (3) the spray dryer. The discussion is limited to the heat, mass, and momentum aspects of simulation and does not consider product quality.

2. SINGLE PARTICLE MODELS

In order to model the dehydration of a food product in a particular dryer, the drying rate characteristics of the individual particles have to be known in terms of the moisture and temperature changes occurring at different drying conditions. The drying rate equation fulfills this purpose. In the simulation of a particular dryer, the drying rate equation is an integral part for the multi-equation dryer model.

To solve the drying rate equation of a food product, the boundary condition needs to be specified in terms of the moisture sorption isotherm which expresses the average moisture content $\bar{M}$ as a function of the water activity and the product temperature. The drying rate and the moisture sorption characteristics are product dependent and are obtained in small-scale laboratory experiments.

Food products can be classified into three categories with respect to drying behavior: (1) liquid solutions (e.g. milk, fruit juices), (2) capillary-porous rigid particles (e.g. grains, wood), and (3) colloidal materials (e.g. meat, vegetables). The sorption and drying rate characteristics of these three food groups have been the subject of extensive scientific investigations (Mujumdar, 1980, 1983, 1984). Although many of the published results are of fundamental interest, this reviewer will limit his review to those relationships of the sorption isotherm and drying rate equations which have proven to be of practical significance to engineers in the design of food dehydration equipment.

2.1 Sorption Isotherms

Hundreds of papers have been published on the sorption phenomena of food products in the past twenty-five years (Chirife and Iglesias, 1978; Pixton and Howe, 1983). In fact, publication of such papers appears to continue unabated (Chinnan and Beuchat, 1985). The purpose of this activity is not evident since sufficient information appears to be known on the topic for design.

A consensus has developed recently among food scientists and engineers of the superiority of the Guggenheim-Anderson-de Boer (GAB) equation for the prediction of the water activity of food products (van den Berg, 1984). The GAB equation has the following form:

$$\bar{M} = \frac{\bar{M}_o C K}{(1 - K a_w)(1 - K a_w + C K a_w)} \qquad [1]$$

where $\bar{M}_o$ is the moisture content at monolayer coverage; and C and K are empirical temperature dependent product constants.

Lomauro et al. (1985a, 1985b) recently compared several sorption isotherm equations for different grains, fruits, vegetables and meats. They concluded that the GAB equation gives a good fit for over 75 percent of the food products studied, and describes the food isotherms better than the two parameter equations. The values of $\bar{M}_o$, C and K in the GAB equation for 163 foods are tabulated in the Lomauro papers for a limited temperature range.

For grains the two-parameter Chung-Pfost isotherm equation has proven to be effective in the simulation of different dryer designs (Fortes and Okos, 1980):

$$\ln a_w = - \frac{K_1}{R\, T_{abs}} \exp(- K_2 \bar{M}) \qquad [2]$$

where K_1 and K_2 are product constants. Values of the two constants for the major grain species over the full range of temperatures encountered in the grain drying processes have been tabulated by Bakker-Arkema et al. (1978).

The GAB equation has a theoretical basis, and thus supplies fundamental information on the thermodynamic properties (e.g. heat of vaporization, pore structure) of a food product. In contrast, the Chung-Pfost relationship is empirical in nature. Which equation is better for design depends on which relationship represents the experimental data best within the range of moisture contents and temperatures encountered in the dryer under study.

2.2 Internal Moisture and Heat Transfer

As with the topic of sorption isotherms, hundreds of papers have been published in the past decades on the topic of single particle or droplet mass of non-food and food products. However, few are relevant to the topic of engineering design. Since several exhaustive reviews on the fundamental aspects of the drying of food materials have recently been published (van Brakel, 1980; Bruin and Luyben, 1980; Karel and Flink, 1983; Toei, 1983), this author will limit his contribution to a review of those equations which have proven to be effective in the actual design calculations of food dryers.

The drying of individual food particles is due to various transport mechanisms. Moisture movement can be caused by liquid diffusion, vapor diffusion, thermal diffusion, capillary flow, internal evaporation and condensation, surface diffusion, shrinkage pressure, or most frequently a combination of these mechanisms. Since knowledge of the exact mechanisms in food dehydration is still limited, the choice of one mechanism (and thus one equation) over another (and thus another equation) for the interpretation of the drying rate data appears nonsensical to the design engineer. What he desires is an equation that represents truthfully the moisture and heat transfer behavior of the food particles in a certain temperature/humidity range.

The drying theory of individual particles has developed from the empirical drying rate equations of Fisher (1923) to the liquid diffusion equation of Gilliland and Sherwood (1933), the capillary movement equation of Ceaglske and Hougen (1937), the vapor diffusion - capillary liquid movement equation of Krischer (1938), the evaporation-condensation theory of Henry (1939), and finally the irreversible thermodynamic equations of Luikov (1966). Recent theoretical drying papers by such authors as Fortes and Okos (1980) and Whitaker (1980) appear to offer nothing fundamentally new on the subject. The best critique on the plethora of so-called "fundamental" papers on the topic of mass transfer in convective drying of porous media is by van Brakel (1980).

Due to the complexity of the drying process, engineers prefer to lump the effects of the different drying transport mechanisms into a simplified moisture diffusion equation with a concentration and temperature dependent diffusion coefficient (D). For a spherical non-shrinking particle:

$$\frac{\partial M}{\partial t} = \frac{1}{r^2} \frac{\partial}{\partial r} \left(r^2 D \frac{\partial M}{\partial r}\right) \qquad [3]$$

The D-value is the apparent or effective diffusion coefficient with an Arrhenius type temperature dependency

$$D = D_o \exp\left(- \frac{E}{R\, T_{abs}}\right) \qquad [4]$$

where E is the energy of activation. Values for D_o and E for some capillary-porous and colloidal food materials at different moisture contents (e.g. wheat, apples, potatoes) have been tabulated by Bruin and Luyben (1980). Unfortunately, knowledge of the precise dependence of the moisture diffusivity on temperature and moisture content of food products is limited, and this often forces engineers to employ empirical drying rate equations for the simulation models of food dryers.

The drying rate of cereals is frequently expressed in the literature by a semi-empirical relationship, the so-called "thin-layer" equation, which relates the standardized average moisture content to time:

$$\frac{\overline{M} - \overline{M}_e}{\overline{M}_i - \overline{M}_e} = \exp\,(-\,k^1 t) \qquad [5]$$

where k^1 is a product constant dependent on temperature and is called the "drying constant" with dimension equal to s^{-1}. (k^1 can thus be interpreted as the ratio of the diffusion coefficient divided by a characteristic length squared.) Values of k^1 for different grains can be found in Brooker et al. (1974).

In comparing equations [3] and [5] for engineering application, the diffusion equation [3] is preferred since it allows calculation of the moisture profiles within a grain kernel. (Equation 5 only calculates the average moisture content.) Also, the diffusion equation simulates the change in the moisture distribution within the grain kernels during periods of tempering in multi-stage packed-bed dryers and in cascading rotary dryers.

In the previous paragraphs of this section, the drying of capillary-porous and colloidal solid particles was considered. In spray dryers liquid foods are dried. The topic of food-droplet drying has been the topic of extensive investigations by Thijssen, Bruin and co-workers in the Netherlands (Thijssen and Rulkens, 1968; Kerkhof and Thijssen, 1977; Bruin and Luyben, 1980; Luyben and Bruin, 1980). The basic diffusion equation [3] was solved numerically by van der Lijn (1976) for droplets changing in volume during the drying process due to contraction or expansion. An approximation to van der Lijn's solution for the case of a power-function variation of the diffusivity with moisture content was developed by Liou (1982) for shrinking spherical food particles.

It appears that equation [3] with the van der Lijn-Liou modifications adequately represents the moisture transport inside liquid foods for engineering applicaton in the calculations of spraydryer design. A requirement for its successful employment is knowledge of the effect of moisture content and temperature on the diffusion coefficient for the particular liquid food to be dried.

It is usually assumed that the resistance to heat transfer during the drying of a food particle or slurry is located in the boundary layer due to the much larger value of the thermal diffusivity than of the effective moisture diffusion coefficient (i.e. $\alpha \gg D$). Hence, the Biot heat transfer coefficient is considered to approach zero in food dryer analyses. This implies that the temperature gradients in food products are close to zero during the drying process (van der Lijn, 1976).

2.3 External Heat and Moisture Transfer

In the majority of commercial dryers, the energy for evaporation is conveyed to the drying food particles by a stream of heated air. The airstream also carries away the water vapor from the surfaces of the particles. The relative velocity between the particles and the airstream determines the values of both the convective heat transfer and convective moisture transfer coefficients. These two parameters are usually expressed in terms of dimensionless numbers; the dimensionless heat transfer coefficient (the Nusselt number) in terms of the Reynolds and Prandtl numbers, and the dimensionless moisture transfer coefficient (the Sherwood number) in terms of (the Reynolds and Schmidt number). It is the task of the engineer to select the appropriate functions for a particular dryer type.

For single food particles moving relative to air as in spraydryer, the heat and moisture transfer coefficients can be calculated from (Ranz and Marshall, 1952):

$$Nu = 2.0 + 0.6\, Re^{0.5}\, Pr^{0.5} \qquad [6]$$

$$Sh = 2.0 + 0.6\, Re^{0.5}\, Sc^{0.33} \qquad [7]$$

For a packed-bed of food particles the convective heat transfer coefficient can be obtained from (Boyce, 1966):

$$Nu = \frac{C_1 d}{k} \left\{ \frac{\varepsilon\, \rho_a\, V_a\, (T + C_2)}{P_{at}} \right\}^{C_3} \qquad [8]$$

where the C- values are dependent on the type of food particles and type of bed (Parry, 1985). The value of the convective mass transfer of a packed-bed of food particles is seldom required in packed-bed simulation models (see section 3.1). This is due to the fact that the major resistance to moisture transfer from solid food particles is located within the particles.

For rotary dryers no universally accepted relationship could be found in the literature for the convective heat transfer coefficient. There appears to be a great deal of disagreement about the value of the h-value among researchers due to the uncertainty of the total surface area of the particles in contact with the drying air at any instant (Baker, 1983). Schofield and Glikin (1962) recommended the following expression:

$$Nu = 0.33\, Re^{0.6} \qquad [9]$$

As is the case with packed-bed dryers, the convective mass transfer coefficient can be assumed to be infinite for the rotary drying of food particles.

The accuracy of the calculated heat and moisture transfer coefficients in a dryer model is limited due to the unknowns in the fluid field and the particle shape. However, the errors in the values of the Nu and Sh numbers appear acceptable compared to other inaccuracies in the food dryer models.

3. DRYER MODELS

Three dryer models are described in this section. The simulation model of the concurrent-flow packed-bed (CCFPB) dryer is available in the open literature, and has been successfully employed for commercial design. The model of the concurrent-flow cascading-rotary (CCFCR) dryer has been developed; it is suitable for industrial applicaton. The simulation model of the concurrent-flow vertical (CCFVS) dryer is in an advanced state of development, and is expected to become available soon as a design tool.

3.1 Packed-Bed Dryer

A packed-bed food dryer consists of the combination of a drying chamber filled with food particles, and a fan/heater assembly. In a concurrent-flow packed-bed-dryer, the particles and the air flow through the chamber in the same direction. The velocity of the bed particles is slow compared to that of the air. Commercial CCFPB dryers in the food industry are multi-stage units, with a tempering zone separating the individual drying stages. A counterflow cooler sometimes complements the CCFPB dryer. Figure 1 is a schematic of a two-stage CCFPB dryer employed for the drying of parboiled rice, with air-recycle of the cooling and second-stage drying air.

To model the CCFPB dryer, heat and mass balances are written on an arbitrarily selected control volume in the dryer, resulting in differential equations for the air temperature (T), the product temperature (θ), the air specific humidity (H), the average product particle moisture content ($\overline{M}$), and the moisture content in the product particles (M), as a

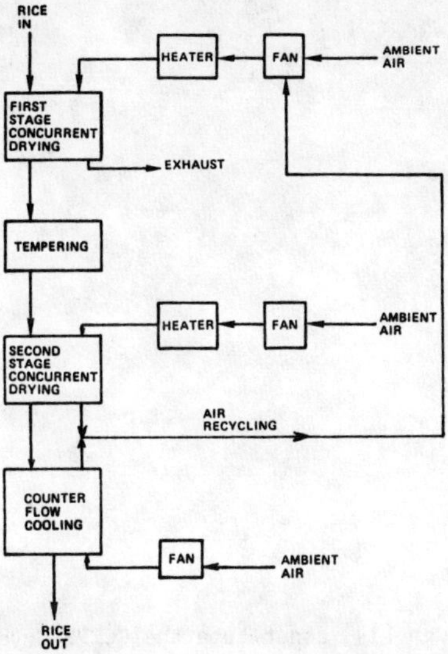

Fig. 1. Schematic of a two-stage concurrent-flow packed-bed (CCFPB) dryer.

function of location in the dryer. Since the flow of the individual particles and the air is uniform over the cross-sectional area of the dryer, the modeling of the air and particle velocity in the dryer is trivial.

The steady-state model for the CCFPB dryer of food particles was first presented by Bakker-Arkema et al. (1971) and has recently been reviewed by Sharp (1985). For spherical food particles, the CCFPB model consists of the following five differential equations:

$$\frac{dT}{dx} = \frac{-ha}{G_a c_a + G_a c_v H}(T - \theta) \qquad [10]$$

$$\frac{d\theta}{dx} = \frac{ha}{G_p c_p + G_p c_w \bar{M}}(T - \theta) - \frac{h_{fg} + c_v (T - \theta)}{G_p c_p + G_p c_w \bar{M}} G\, C_a \frac{dH}{dx} \qquad [11]$$

$$\frac{dH}{dx} = - \frac{G_p}{G_a} \frac{d\overline{M}}{dx} \qquad [12]$$

$$\frac{\partial M}{\partial t} = \frac{1}{r^2} \frac{\partial}{\partial r} \left(Dr^2 \frac{\partial M}{\partial r}\right) \qquad [13]$$

$$\overline{M} = \frac{1}{R} \int_0^R M \, dr \qquad [14]$$

Equations [10] through [14] constitute the CCFPB dryer simulation model. Equations [13] and [14] are the single-particle drying rate equations for the moisture content distribution within the particles and the average moisture content, respectively. Numerical solution of the model is by a standard Runga-Kutta integration technique (Bakker-Arkema et al., 1974).

The simulation of a tempering stage in a multi-stage packed-bed concurrent-flow dryer is accomplished by solving eqn. [14] for isothermal and isomoisture conditions over a time-period equal to the traverse-time of the food particles through the tempering zone.

3.2 Cascading Rotary Dryer

A cascading rotary dryer consists of a slightly inclined long cylindrical shell which rotates slowly. In a concurrent-flow cascading rotary dryer, the moist particles and hot drying air are fed into one end of the dryer and the dried (or partially dried) particles along with the moist air exit at the other end. The inside of the shell is fitted with a set of flights which lift the particles partway along the periphery of the rotating shell before they fall back onto the bottom of the dryer where they are picked up once more to go through the same process. As the particles fall from the top to the bottom of the shell, they encounter the flow of hot air which moves at considerably higher speed than the food particles. Figure 2 shows a CCFCR dryer employed for the drying of parboiled rice. Modeling of the cascading rotary dryer requires the simultaneous solution of a series of equations expressing (i) the heat and mass transfer of the individual particles, and (ii) the movement of the particles in the cascading rotary dryer.

By making heat and mass balances on a cross-section of a CCFCR dryer, the same equations result as for the concurrent-flow packed-bed dryer

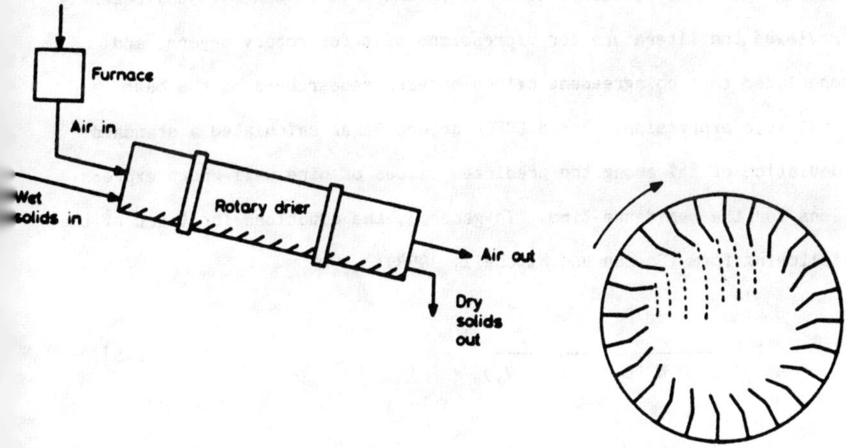

Fig. 2. A concurrent-flow cascading rotary (CCFCR) dryer.

(see equations [10] through [14]). In fact, Sharples et al. (1964) employed a steady-state rotary dryer model which is very similar to the CCFPB model of eqns. [10] through [14].

The major complicating difference in the simulation models of the packed-bed and the rotary dryers is found in the term G_p, the particle transport rate through the dryers. In the case of the packed-bed dryer, G_p is a direct function of the positive auger-displacement; for the rotary dryer the particle transport is a complicated function of the cascading, bouncing, and rolling actions encountered by the particles in the dryer.

The term G_p, the effective particle flow rate across the cross-sectional area of a CCFCR dryer, is equal to the product of the product density (ρ_p) and the effective velocity (V_p) of the particles along the dryer axis. In turn, V_p is the ratio of the dryer length (L) and the average particle residence-time (t) in the dryer. Baker (1985) recently reviewed the literature for expressions of t for rotary dryers, and concluded that no agreement exists between researchers on the best available expression. For a CCFCR dryer, Baker calculated a standard deviation of 86% among the predicted values of nine well-known expressions for the residence-time. In general, the equations for t are of the following form (Saeman and Mitchell, 1954):

$$t = \frac{L}{C_1 \, D \, N \, (\tan \alpha + C_2 \, V_a)} \qquad [15]$$

where L is the dryer length, D the effective inside dryer diameter, N the rotational dryer speed, α the drum slope, V_a the air velocity, and C_1 and C_2 constants depending on the flight design and the food material to be dried.

Numerical solution of the CCFCR dryer model, eqns. [10] through [14], can be obtained with a standard Runga-Kutta integration technique coupled to a finite difference solution of eqn. [9] (Bakker-Arkema et al., 1974).

3.3 Vertical Spray Dryer

A vertical spray dryer consists of a drying chamber, a liquid feed atomizer, and a dry powder removal mechanism. In a concurrent-flow vertical spray dryer, the atomized droplets and the air both flow in the same vertical downward direction. The droplet and air velocities are of the same order of magnitude. Figure 3 shows the schematic of a CCFVS dryer.

Modeling of the spray drying process requires the knowledge of the

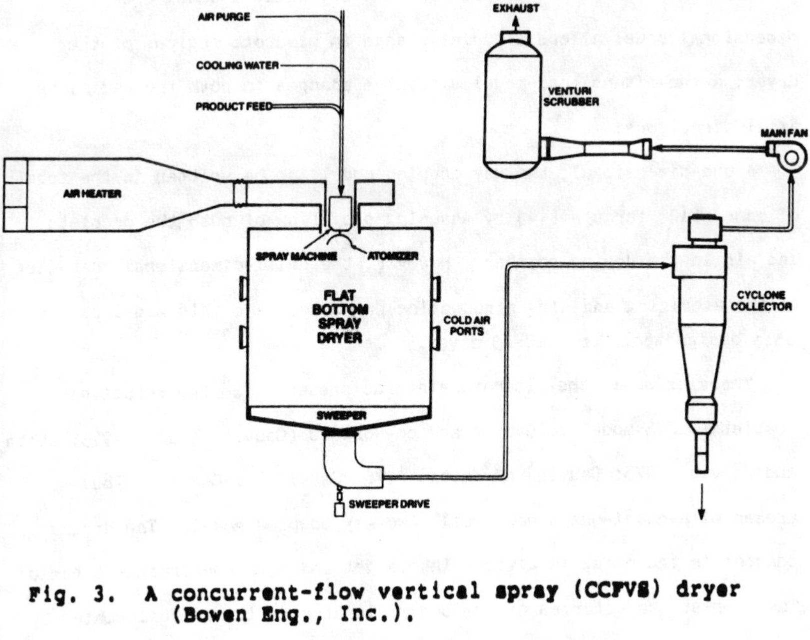

Fig. 3. A concurrent-flow vertical spray (CCFVS) dryer (Bowen Eng., Inc.).

droplet size distribution, mass fraction and initial velocity. The CCFVS dryer model consists of equations for: (i) the three-dimensional motion of the droplets, (ii) the three-dimensional flow pattern of the drying gas, and (iii) the heat and mass transfer of the cloud of droplets. For the packed-bed dryer, the solution of motion of the particles and the air is trivial, while for the rotary dryer only the motion of the solid particles requires special attention (Equation [15]). However, in the spray dryer model the equations of motion for both the droplets and the drying medium need to be solved, not just in one but in three dimensions.

Crowe (1980) classified the models presented in the literature in terms of coupling and geometry. One-way coupled models only account for the effect of the drying air on the droplets; two-way coupling includes the effect of the air on the droplets as well as the effect of the droplets on the air. A true one-dimensional model accounts only for axial changes in air and droplet properties, while a quasi one-dimensional model allows a radial change in discrete regions of the dryer; a two-dimensional model simulates changes in both the radial and axial directions.

A one-dimensional, two-way coupled model can be written in the form of eqns. [10] through [14] by assuming plug-flow of both the droplets and air in the drying chamber. However, the multi-dimensional character of the atomizing and air-introduction processes make this model useless as a design tool for a CCFVS dryer.

There exist in the literature at the present time two relatively complete CCFVS models. Gauvin and co-workers (Gauvin et al., 1975; Katta and Gauvin, 1975; Gauvin and Katta, 1976; Gauvin and Costin, 1980) presented a quasi-one-dimensional, two-way coupled model. The drying chamber in the model is divided into a jet and a free-entrainment region. The droplet trajectories of the pneumatic atomization are calculated by integrating the equation of motion in centrifugal and gravitational fields (Coulson and Richardson, 1968). This leads to expressions for the tangential (t), axial (v) and radial (r) velocities (V) of the droplets, and thus to the resultant velocity (V_f) of a droplet relative to the air (a):

$$V^2_f = (V_t - V_{at})^2 = (V_v - V_{av})^2 = (V_r - V_{ar})^2 \qquad [16]$$

The air flow-field (V_{at}, V_{av}, V_{ar}) is assumed to be equal to the superposition of the atomizing fluid and the surrounding drying-air flow field, and to be unaffected by the droplets. The air temperatures at different axial positions within the nozzle and the entrainment zones are estimated from the basic energy balance equation. The dimensions of the jet and free-entrainment zones are established empirically. The heat and mass transfer to and from the cloud of droplets is calculated from:

$$\frac{dq}{dt} = k_a \, Nu \, (T - \theta) n_i \, d_i \qquad [17]$$

$$\frac{dm}{dt} = D \, \rho_a \, Sh \, (H_{sat} - H) n_i \, d_i \qquad [18]$$

where n_i represents the number of droplets with diameter d_i; k_a and ρ_a are the thermal conductivity and density of the drying air, respectively; H_{sat} is the saturation at the droplet temperature θ; and Nu and Sh are the Nusselt and Sherwood numbers calculated with equations [6] and [7]. Since the droplet diameter of a liquid food changes as the droplet loses moisture, the instantaneous diameter of the particles (assuming the droplets are dense) is:

$$d_i = d_{in} \, (\rho_{in} \, C_{in} / \rho_i \, C_i) \qquad [19]$$

where the subscripts i and in represent the instantaneous and initial values of the diameter droplet (d) and droplet solid concentration (C), respectively.

The Gauvin model requires that the dimensions and the radial distribution of axial velocities of the entrainment zone are known. Frequently this is not the case, and for this reason the two-way coupled, two-dimensional model of Crowe (1980) is of a more general nature.

In the Crowe (1980) CCFVS dryer model, the droplets are treated as sources of moisture, momentum and energy to the drying air phase. The flow field in the dryer is divided into a network of two-dimensional cells. Droplet trajectories, and the moisture and temperature along each trajectory, are calculated throughout the cell network. The drying air flow field is originally calculated assuming no droplets are present, and is subsequently recalculated using the previously determined droplet effects. Repeated integration converges the solution to the true properties of the droplet and drying air in every cell throughout the spray dryer. The method is called the PSI-call method. The reader is referred for specifics of the method to the original paper on the topic by Crowe et al. (1977), and to two more recent applicatons of the technique by Arnason and Crowe (1980) and by Crowe (1984).

The basic equations for droplet heat and mass transfer equations [6], [7], [17] and [18] are the same for the Gauvin and Crowe models.

4. DRYER MODELING

Limited results of the application of the three dryer models are presented in this section. The drying of rice in a concurrent-flow packed-bed dryer is examined, followed by the drying of fertilizer in a concurrent-flow rotary dryer and the evaporation of water in a concurrent-flow vertical spray dryer. Equipment sizing, sensitivity analysis, and optimum design are among the topics considered.

Table 1. Experimental drying data of medium-grain rice dried in a three-stage concurrent-flow packed-bed dryer.

Test No.	Temperature Setting (°C) and Static Pressure (kPa)			Rice Velocity (m/hr)	Average Moisture Content (% Wet Basis)		White Rice Head Yield (%)		Dryer Efficiency (kJ/kg)
	Top Stage Temp (SP)	Middle Stage Temp (SP)	Bottom Stage Temp (SP)		In	Out	In	Out	
1	177(3.5)	115(3.5)	88(3.5)	4.9	22.1	16.6	50.5	49.6	4287
2	149(2.0)	107(3.5)	71(3.5)	3.7	23.5	17.5	43.2	47.6	3052
3	143(3.0)	121(3.7)	65(4.2)	3.9	18.5	13.6	53.1	49.3	3493
4	135(2.7)	107(3.2)	82(2.5)	3.9	25.5	19.9	48.3	49.7	3456
5	121(1.2)	110(2.7)	104(2.7)	4.9	22.4	18.5	49.0	50.9	3963
6	118(1.5)	110(2.7)	82(2.7)	4.0	22.6	18.9	40.8	47.7	3012
7	99(1.4)	104(2.9)	104(2.9)	4.8	21.0	16.9	50.5	49.2	3298
8	93(1.5)	88(3.0)	82(2.7)	3.9	18.7	14.6	45.8	47.0	4249

from Bakker-Arkema et al., 1983

Note: The units in the Tables and Figures are the same as in the original references.

4.1 Packed-Bed Dryer

Rough rice can be dried in concurrent-flow packed-bed dryers at high inlet air temperatures without affecting the rice quality. Typical experimental data of a three-stage CCFPB dryer is shown in Table 1; the inlet air temperature ranged from 65-175°C (Bakker-Arkema et al., 1983). Note that the rice quality (e.g. head yield) did not deteriorate notwithstanding the high drying air temperatures.

Modeling has been applied extensively in the design of commercial CCFPB dryers for food products. In the next paragraphs some of the simulation results obtained with the CCFPB model described in Section 3.1 are discussed.

Application of the CCFPB model to a particular food product such as rice, requires knowledge of the relevant product parameters discussed in section 2 (e.g. the sorption characteristics, the effective diffusion coefficient, the heat of vaporization, etc.). For rice these quantities

Table 2. Simulationresults of the first stage of a three-stage concurrent-flow packed-bed dryer drying medium-grain rice at 135°C at a grain velocity of 3.8 m/h and airflow rate of 1800 m/h.

Depth m	Time hr	Air Temp C	Abs Hum kg/kg	Rel Hum dec.	Grain Temp* C	MC %, w.b.
0.00	0.00	135.0	.0060	.003	23.9	25.5
0.15	.04	43.9	.0214	.360	43.9	24.3
0.30	.08	39.8	.0283	.586	39.8	24.1
0.46	.13	38.0	.0312	.710	38.0	24.0
0.62	.18	36.9	.0331	.796	36.9	23.9
0.76	.22	36.3	.0341	.846	36.3	23.9
0.91	.26	35.6	.0352	.911	35.6	23.9
1.07	.31	35.5	.0354	.917	35.5	23.8

* maximum rice temp = 51.5°C at a depth of 0.0155 m.

Bakker-Arkema et al., 1983

have been determined to an accuracy adequate for engineering design (Bakshi and Singh, 1979; Steffe and Singh, 1980; Bhattacharya and Ali, 1985; Zuritz and Singh, 1985). Validation of the CCFPB model preceded its use as a tool for engineering design (Fontana, 1983).

The simulated moisture content, air temperature, and rice temperature in a three-stage CCFPB dryer is illustrated in Figure 4. Similar information in table form is contained in Tables 2 and 3. Figure 4a and 4b furnish the engineer with an overall picture of the operation of the dryer, and provides insight on for instance, the approximate moisture removed in each stage. More specific knowledge about the maximum rice temperature and the moisture extraction (both parameters affecting the rice quality) can be obtained from Table 2. The data in Table 3 on the internal moisture content distribution in each rice kernel during the tempering process between consecutive drying treatments allows the designer to calculate the length of the tempering zone.

Table 3. Simulated internal moisture content distribution (% d.b.) in a medium-grain rice kernel after leaving the first stage of a three-stage concurrent-flow packed-bed dryer and tempering for 1.35 hours.

Time hr	Distance from center of kernel, m						
	Center node						Surface node
	0.00000	0.00033	0.00056	0.00078	0.00116	0.00122	0.00156
0.00	34.3	34.2	33.7	31.5	30.1	26.8	22.4
1.35	32.1	31.9	31.6	31.2	31.1	30.9	30.7

from Bakker-Arkema et al., 1983

The simulated results presented in Figs. 4a and 4b and Tables 2 and 3 have formed the basis for the practical design of two CCFPB rice dryers for the commercial rice-milling and rice-parboiling industry; the field performance of both units closely approximates the simulated performance

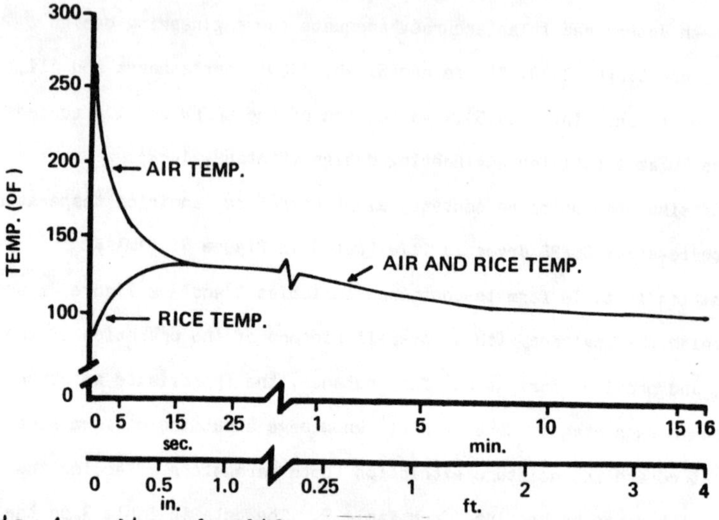

Fig. 4a. Air and solids temperature distribution in a one-stage CCFPB dryer (Brooker et al., 1984).

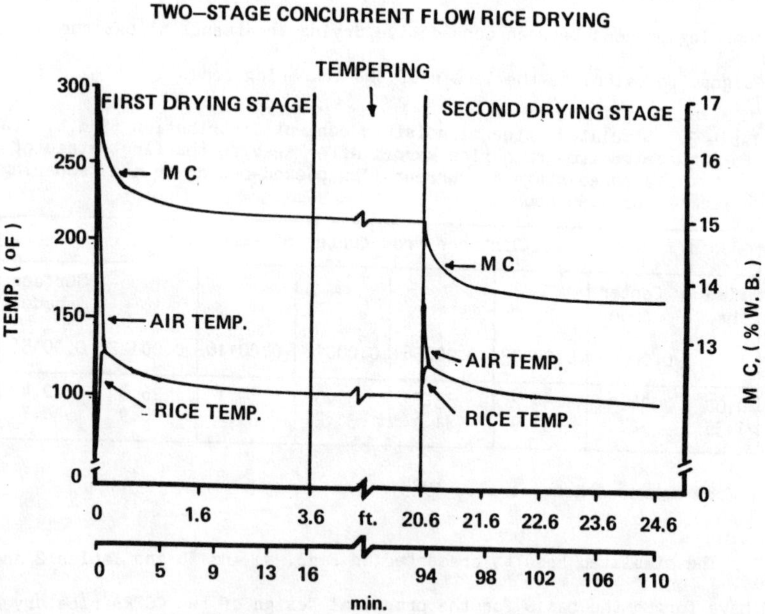

Fig. 4b. Air and solids temperature and moisture-content profiles in a two-stage CCFPB dryer (Bakker-Arkema et al., 1984).

with respect to capacity, moisture extraction, and the drying air temperature/air flowrate effects (Bakker-Arkema et al., 1983; Bakker-Arkema et al., 1984).

The CCFPB dryer simulation model has recently been applied successfully to the commercial design of soybean dryers in oil extraction plants (Bakker-Arkema et al., 1983), and of sorghum dryers in starch manufacturing plants (Fedewa, 1985). In both cases the CCFPB simulation model greatly speeded up dryer analyses, saved engineering-time, and resulted in greater thermal efficiency of the final design (Westelaken, 1985).

Brook and Bakker-Arkema (1980) extended the use of the CCFPB model, beyond straightforward application of eqns. [10] through [14], to the area of optimization. Dynamic programming was used to find the optimum values of the operational parameters and the dryer dimensions; the objective function was based on the fixed and the energy costs. The optimization scheme has not yet been applied to commercial design due mainly to the sensitivity of the factors in the objective function and to the lack of an acceptable quality constraint equation for most food products. However, the author expects that these shortcomings in the application of optimization techniques in commercial CCFPB dryer design will soon be overcome.

4.2 Cascading Rotary Dryer

After parboiled rice is soaked and steamed, it is too sticky to dry in a packed-bed dryer. Rotary dryers are often used for this purpose (Gariboldi, 1974). Typical experimental data of three CCFCR units in-series drying parboiled rice in one pass from 35 to 14 percent (wet basis), is shown in Table 4; the air temperatures varied from 288°C in the first dryer, to 232°C in the second, to 149°C in the last unit (Bakker-Arkema et al., 1984). Drying data of parboiled rice dried in a combination of a

CCFCR dryer and a CCFPB dryer in-series is shown in Table 5. Note that the fuel consumption of the CCFCR/CCFPB dryer combination exceeds that of the three in-series CCFCR units.

Table 4. Experimental data of three-rotary dryer system in a commercial rice parboiling plant.

Rotary Dryer No.	Inlet MC % w.b.	Outlet MC % w.b.	Water Evap. kg/hr	Fuel Cons. m^3 Nat. gas/hr	Fuel Eff. kJ/kg
1	34.6	23.3	2,173	239	4,105
2	23.3	17.8	841	166	7,408
3	17.8	14.1	507	119	8,729
2 + 3	23.3	14.1	1,348	285	7,867
Total System	34.6	14.1	3,521	524	5,545

from Bakker-Arkema et al., 1984

Table 5. Experimental data of one rotary dryer /one concurrent-flow packed-bed dryer system in a rice parboiling plant.

Dryer Type	Inlet MC % w.b.	Outlet MC % w.b.	Water Evap. kg/hr	Fuel Cons. m^3 Nat. gas/hr	Fuel Eff. kJ/kg
Rotary	34.6	24.1	2,044	225	4,105
CCFPB	24.1	15.5	1,290	146	4,203
Total System	34.6	15.5	3,334	371	4,143

from Bakker-Arkema et al., 1984

No modeling studies have as yet been published on rotary drying of food particles. However, the model of the CCFCR dryer described in Section 3.2 has been used to analyze the drying of granular fertilizer in 8.5 diameter, 54 ft long dryer with 24 flights (Sharples et al., 1964). The simulation results of the Sharples study are considered in the next paragraphs.

Sharples et al. (1964) obtained values of the relevant fertilizer drying properties as well as of the equipment heat-transfer coefficient from the literature. In the model Equations [10] through [14], the solids velocity was calculated from an expression similar to Equation [7] developed by Schofield and Glikin (1962). No validation of the CCFCR model was attempted.

The simulated moisture content, air temperature and product moisture content in the CCFCR dryer are illustrated in Figure 5 and gives the designer an excellent overview of the operation of the CCFCR. Location and level of the maximum particle temperature are shown along with the

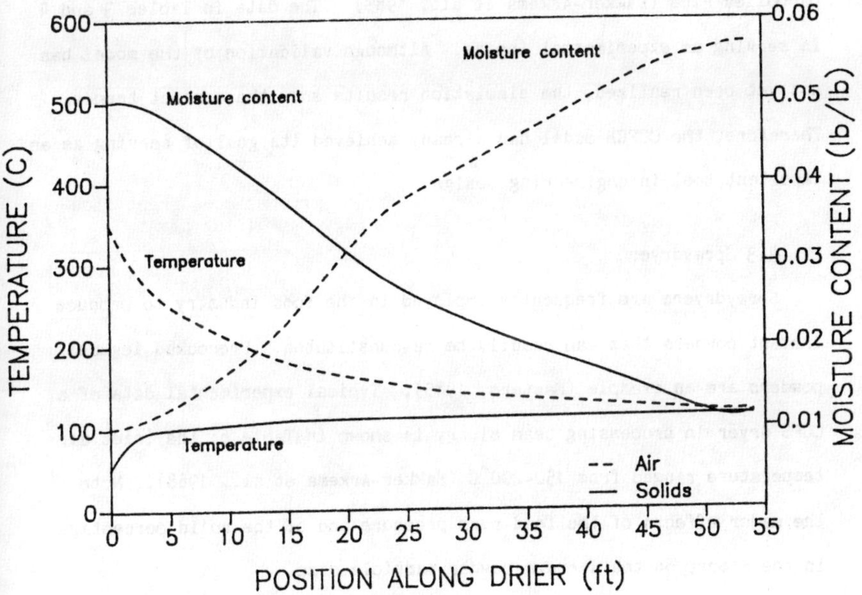

Solids rate= 1500 lb/min (40.2 ton/h)
Air rate= 1200 lb/min

Fig. 5.–Air and solids temperature and moisture–content profiles in a CCFCR dryer (Sharples et al.,1964).

decrease in moisture content. The effect of several operating parameters (e.g. drying air temperature, airflow rate, speed of rotation, product flowrate) on the outlet moisture content is illustrated in Figures 6 and 7, and allows the engineer to quickly make an approximation of the influence of the main operating parameters on the drying process.

The fact that the CCFCR dryer simulation model has not been validated, has obviously prevented its use in industry (Purcell, 1979). Lack of a reliable transport model prevents others from employing simulation in the design of CCFCR dryers (Reay, 1979).

The author is presently testing the Sharples CCFCR dryer model for parboiled rice (Bakker-Arkema et al., 1985). The data in Tables 3 and 4 is serving as experimental inputs. Although validation of the model has not yet been realized, the simulation results show the correct trends. Therefore, the CCFCR model has already achieved its goal of serving as an important tool in engineering design.

4.3 Spraydryer

Spraydryers are frequently employed in the food industry to produce instant powders that can readily be re-constituted. Precooked legumes powders are an example (Masters, 1985). Typical experimental data of a CCFS dryer in processing bean slurry is shown in Table 6; the inlet air temperature ranged from 150-200°C (Bakker-Arkema et al., 1968). Note the major effects of the feed pump pressure and of the solid percentage in the slurry on the average powder particle size.

Spraydryers in the food industry are usually of the concurrent-flow type in which the liquid feed and drying gas flow through the drying chamber in the same direction. The task of the engineer is to design the optimum drying conditions and dryer dimensions for drying the liquid particles adequately before they contact the walls of the drying chamber.

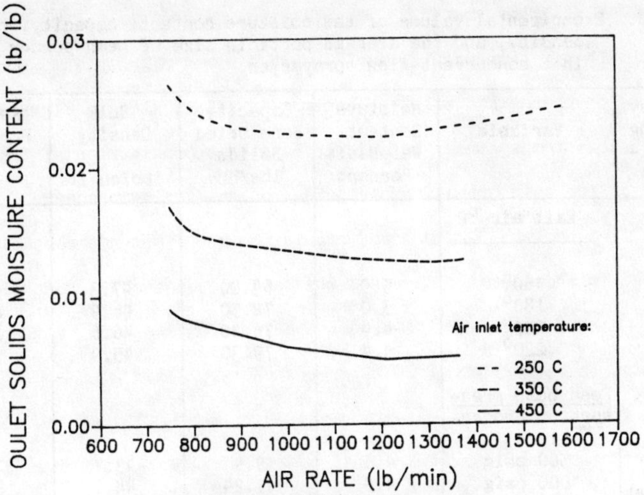

Solids rate= 1500 lb/min
Inlet solids moisture content= 0.05 lb/lb

Fig. 6.–Effect of air rate and inlet air temperature in a CCFCR dryer (Sharples et al.,1964).

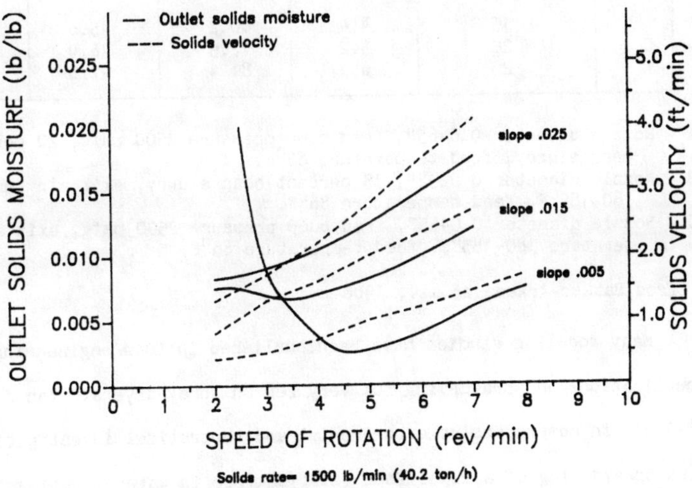

Solids rate= 1500 lb/min (40.2 ton/h)
Air rate= 800 lb/min
Inlet solids moisture content= .05 lb/lb
Inlet air temperature= 400 C
Inlet solids temperature= 80 C

Fig. 7.–Effect of varying slope and speed of rotation on drying and solids velocity in a CCFCR dryer (Sharples et al.,1964).

Table 6. Experimental values of the moisture content, capacity, bulk density, and the average particle size of bean powder dried in a concurrent-flow spraydryer.

Spray Drying Conditions	Variable	Moisture Content Wet Basis Percent	Capacity Bone Dried Solids lbs/hr	Bulk Density lbs/cu ft	Average Particle Size (μ)
	Exit air °F				
A	160°	5.9	63.00	47.3	93
	180°	5.0	72.50	46.9	82.4
	200°	4.9	76.40	46.6	80.2
	220°	4.4	79.30	45.9	73.3
	Feed pump pressure (unhomogenized)				
B	500 psig	4.5	9.91	43.7	122.4
	1000 psig	4.9	17.24	44.0	88.9
	2000 psig	5.3	26.38	45.9	82.3
	3000 psig	5.2	31.95	45.9	54.3
	Percent solids in bean slurry				
C	15	4.4	60.2	45.6	59.1
	20	5.2	72.5	46.9	82.4
	25	6.1	81.4	47.6	123.8

A - Nozzle diameter 0.0465", feed pump pressure 2500 psig, 20 percent bean slurry, feed temperature 86°F.
B - Nozzle diameter 0.0292", 15 percent bean slurry, exit air temperature 180-185°F, feed temperature 86°F.
C - Nozzle diameter 0.0465", feed pump pressure 2500 psig, exit air temperature 180-185°F, feed temperature 86°F.

from Bakker-Arkema et al., 1968

Many modeling studies have been published in food engineering publications on spraydrying but very few on spray dryers. Van der Lijn (1976), in comprehensive experimental and theoretical investigation of the spraydrying of a model food (i.e. maltose in water), employed a one-dimensional, one-way coupled model in analyzing a concurrent-flow spray dryer; the effect of a central air bubble on the simulated temperature history of the maltose droplets was among the topics evaluated. Daemen

(1984) applied the van der Lijn CCFS dryer model in the calculation of the destruction rates of enzymes and bacteria during the spraydrying of milk; surprisingly good agreement was obtained between the experimental and computed results. Although van der Lijn's spray dryer model has some excellent features, it lacks the sophistication in the droplet trajectory calculations featured in the Gauvin model (Gauvin et al., 1975), and especially the Crowe model (Crowe et al., 1977). Some simulation results of the Crowe model are considered in the next paragraphs.

Arnason and Crowe (Arnason et al., 1980; Crowe, 1980) used the experimental values of Manning and Gauvin (1960) to validate their model; water droplets were evaporated in a 0.2 m diameter CCFS dryer. The droplet size distribution was represented by four discrete sizes ranging from 5 to 40 μm in diameter, and from 9.5 to 80 m/sec in droplet velocity; the starting angle for each trajectory was 7.5^{o}. The initial air temperature was $106^{o}C$, the airflow rate 0.038 kg/s, and the water feed rate 5.5×10^{-4} kg/s.

The simulated average air and droplet temperatures along the dryer axis are shown in Figure 8; the simulated data agree well with the experimental values of Manning and Gauvin (1960). Assuming temperature equilibrium between the droplet and air temperature, as is done in several spraydryer models, is obviously questionable. The simulated radial air velocity distribution is illustrated in Figure 9. The multi-dimensional nature of the distribution is evident, and thus one-dimensional spraydryer models are a gross simplification. Figure 10 shows the radial air temperature distribution; rapid evaporation causes a decrease in the air temperatures especially close to the atomizer within the spray envelope, resulting in the multi-dimensional nature of the temperature distribution. Fig.11 illustrates the typical average air and droplet temperatures in a CCFS dryer.

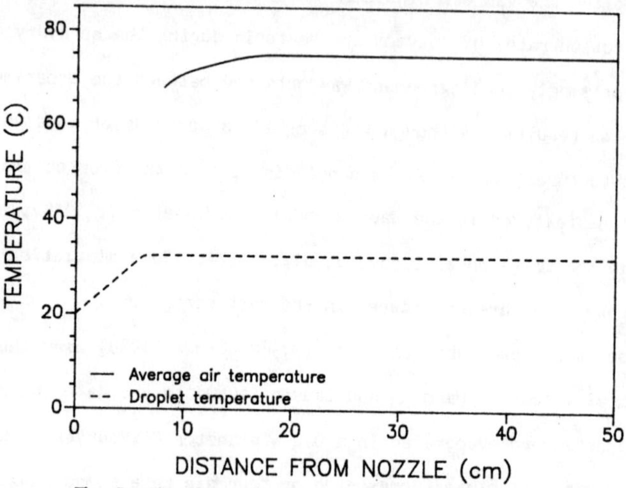

Fig. 8.—Simulated air and droplet temperatures in a CCFVS drier (Crowe,1980).

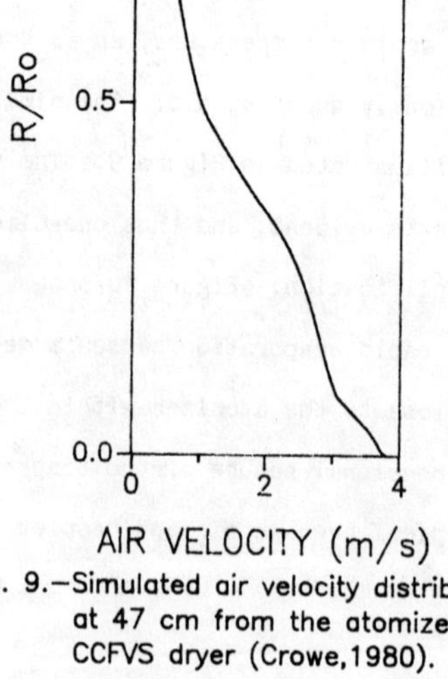

Fig. 9.—Simulated air velocity distribution at 47 cm from the atomizer in a CCFVS dryer (Crowe,1980).

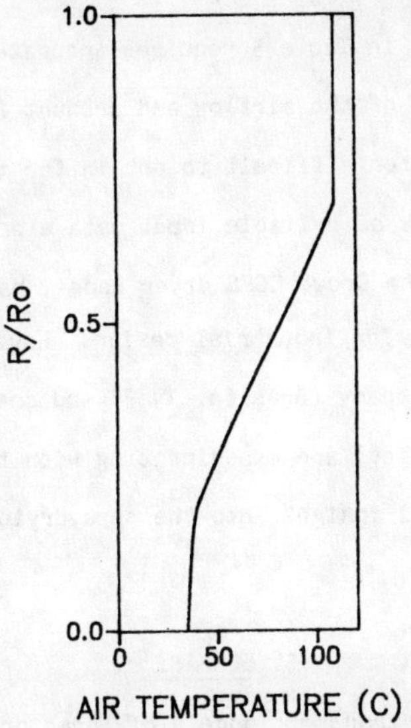

Fig. 10.—Simulated air temperature distribution at 50 cm from the atomizer in a CCFVS dryer (Crowe,1980).

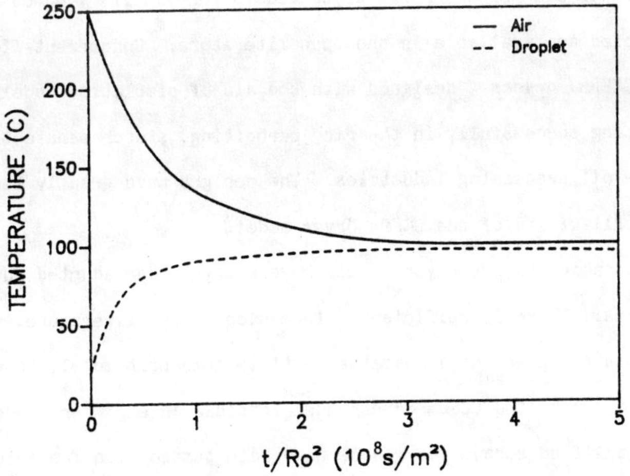

Fig. 11.—Simulated Temperature history of a droplet and its surrounding air in a CCFVS dryer (van der Lijn,1976).

Application of the Crowe CCFS dryer model for evaluation of the bean-slurry dehydration data in Table 5 requires accurate knowledge of the direction and magnitude of the airflow and product flow into the dryer. This type of data is often difficult to obtain for the drying of liquid food products. The lack of reliable input data along with the relatively complicated nature of the Crowe CCFS dryer model, has limited the rapid acceptance of the model for industrial design. However, engineers of at least one major food company (Breslin, 1985) and one major university (Michigan State University) are experimenting with the Crowe model in order to gain additional insight into the spraydrying of liquid food materials.

5. SUMMARY

The heat, mass, and transport models of three dryers used extensively in the food industry have been assessed with respect to their value in engineering design. It was found that the state-of-the-art of modeling has greatly advanced in the past decade but does depend on dryer-type. No longer should dryer design be considered solely an art.

Packed-bed flow-through dryer models for capillary-porous food particles are available in the open literature. Concurrent-flow packed-bed (CCFPB) dryers , designed with the aid of simulation modeling, are operating successfully in the rice parboiling, starch manufacturing, and edible oil processing industries. The designs have greatly benefited from the availability of the CCFPB dryer model.

Cascading rotary dryer models have not yet been adapted for food products. There is sufficient information in the literature, with the possible exception of a suitable particle transport model, to make application of the CCFCR dryer model feasible in engineering design.

Simplified spraydryer models have been tested with food slurries;

their simplified nature prevents use in practical engineering design. Recent advances in spraydryer modeling, especially with respect to the multi-dimensional droplet and air flow rates in the drying chamber, have given hope that a reliable spraydryer model for food products will soon become available to the food engineer.

Heat, moisture, and momentum models of food dryers have to be combined with models of ingredient deterioration [e.g. flavor, vitamins, viability, texture, etc.], in order for simulation modeling to become totally acceptable within the food industry. It is the challenge to the food engineering society to make the combinationof heat, mass, and momentum models with product-quality models a reality.

6. LIST OF SYMBOLS

a	specific product surface area, m^{-1}
a_w	water activity, decimal
c	specific heat, kJ/kg^oC
d	diameter, m
h	convective heat transfer coefficient, $kJ/hr\ m^2\ C$
h_{fg}	heat of vaporization, kJ/kg
k^1	drying constant, sec^{-1}
k	thermal conductivity, W/m^oC
m	meter
P_{at}	atmospheric pressure, N/m^2
r	kernel coordinate, m
t	time, sec
x	depth coordinate, m
C	product constant
D	effective diffusion coefficient, m^2/s
G	dry weight flow rate, $kg/hr\ m^2$

H humidity ratio, kg/kg

K product constant

M local moisture content within kernel, dry basis (decimal)

$\overline{M}$ average moisture content, dry basis (decimal)

Nu Nusselt number, dimensionless

Pr Prandtl number, dimensionless

R equivalent kernel radius, m or gas constant

Re Reynolds number, dimensionless

Sc Schmidt number, dimensionless

T air temperature, ^{o}C

T_{abs} absolute temperature, ^{o}K

V velocity, m/s

θ product temperature, ^{o}C

ε porosity, decimal

ρ density, kg/m^3

as a subscript:

a air

e equilibrium

f resultant

i initial

r radial

t at time t or tangential

v vapor or axial

w water

7. REFERENCES

Arnason, G. and Crow, C.T. 1980. Assessment of numerical models for spray drying. In Drying '80. Mujumdar, A.S. ed. Hemisphere Publishing Corp., New York, NY. p410-416.

Baker, C.G.J. 1983. Cascading rotary dryers. In Advances In Drying. Vol. 2. Mujumdar, A.S. ed. Hemisphere Publishing Corp., New York, NY. p1-51.

Bakker-Arkema, F.W., R.C. Brook and L.E. Lerew. 1978. Cereal grain drying. In Advances In Cereal Science And Technology. Vol II. Pomeranz, Y. ed. AACC, Inc., St. Paul, MN.

Bakker-Arkema, F.W., C. Fontana, R.C. Brook and C.M. Westelaken. 1983. Concurrent-flow rice drying. Drying Technology1(2)171-191.

Bakker-Arkema, F.W., C. Fontana, S.J. Kalchik and C.M. Westelaken. 1984. Drying in commercial rice parboiling plants. Drying Technology 2(3)423-433.

Bakker-Arkema, F.W., S.J. Kalchik, I.P. Schisler and C.M. Westelaken. 1983. Commercial drying of soybeans for processing. Paper No. 83-6527. Am. Soc. Agr. Eng.: St. Joseph, MI.

Bakker-Arkema, F.W., S.J. Kalchik and I.P. Schisler. 1985. Rotary drying of parboiled rice - a simulation model. Drying Technology (to be published).

Bakker-Arkema, F.W., L.E. Lerew, S.F. DeBoer and M.G. Roth. 1974. Grain Dryer Simulation. Research Report 224, Agr. Exp. Sta., Mich. St. Univ., East Lansing, MI.

Bakker-Arkema, F.W., M. Palnitkar and C.L. Bedford. 1968. Spraydrying of pea beans in a concurrent horizontal dryer. Trans. ASAE 11(1) 104-107.

Bakker-Arkema, F.W., J.R. Rosenau and W.H. Clifford. 1971. Measurements of grain surface area and its effects on the heat and mass transfer rates in fixed and moving beds of biological products. Trans. ASAE 14(3)864-867.

Bakshi, A.S., and R.P. Singh. 1979. Drying of high moisture parboiled rice. Paper No. 79-6006. Am. Soc. Agr. Eng., St. Joseph, MI.

van den Berg, C. 1984. Description of water activity of foods for engineering purposes by means of the G.A.B. model of sorption. In Engineering And Foods. Vol. 1. McKenna, B.M. ed. Elsevier Applied Science Publishers, New York, NY. p311-322.

Bhattacharya, K.B. and S.F. Ali. 1985. Changes in rice during parboiling, and properties of parboiled rice. In Advances In Cereal Science and Techn. Vol 7. Pomeranz, Y. ed. AACC, St. Paul, MN. p105-167.

Boyce, D.S. 1966. Heat and moisture transfer in ventilated grain. J. Agric. Eng. Res. 11(4)255-265.

van Brakel, J. 1978. Opinions about selection and design of dryers. In Proc. First International Symposium on Drying. Mujumdar, A.S. ed. Science, Princeton, NJ.

Breslin, J. 1985. Personal Communication. Kellogg Corp., Battle Creek, MI.

Brook, R.C. and F.W. Bakker-Arkema. 1980. Design of multi-stage corn dryers using computer optimization. Trans. ASAE 23(1)200-203.

Brooker, D.B., F.W. Bakker-Arkema and C.W. Hall. 1974. Drying Cereal Grains. AVI Publ. Co., Westport, CT.

Bruin, S. and K. Ch. A.M. Luyben. 1980. Drying of food materials: a review of recent develo pments. In Advances In Drying. Vol 1. Mujumdar, A.S. ed. Hemisphere Publishing Co., New York, NY. p153-215.

Ceaglske, N.H. and O.H. Hougen. 1937. Drying granular solids. Ind. Eng. Chem. 29(6)805-813.

Chinnan, M.S. and L.R. Beuchat. 1985. Sorption isotherms of whole cowpeas and flours. Lebensm. Wiss. U. Technol. 18(2)83-88.

Chirife, J. and H.A. Iglesias. 1978. Equations for fitting water sorption isotherms of foods: Part 1 - a review. J. Food Technol. 13(6)159-174.

Coulson, J.M. and J.F. Richardson. 1968. Chemical Engineering. Vol. 2. Pergamon Press, New York, NY.

Crowe, C.T. 1980. Modeling spray-air contact in spraydrying systems. In Advances In Drying. Vol. 1. Mujumdar, A.S. ed. Hemisphere Publishing Corp., New York, NY. p63-100.

Crowe, C.T. 1984. Droplet - gas interaction in counter-current spray dryers. Drying Techn. 1(1)35-56.

Crowe, C.T., M.P. Sharma and D.E. Stock. 1977. The particle - source-in-cell (PSI-Cell) model for gas-droplet flows. J. Fluids Eng. 99(2)325-332.

Daemen, A.L.H. 1984. The destruction of enzymes and bacteria during spraydrying of milk and whey. Neth. Milk Dairy J. 38(2)55-70.

Fedewa, G.L. 1985. Concurrent-flow drying of grain sorghum and the resulting wet milling quality. MS Thesis. Michigan State University, East Lansing, MI.

Fisher, E.A. 1923. Some moisture relations of colloids. Proc. R. Soc. London A1(103)139-161.

Fontana, C. 1983. Concurrent-flow versus conventional drying of rice. Ph.D Thesis. Michigan State Univ., East Lansing, MI.

Fortes, M and M.R. Okos. 1980. Drying theories: their bases and limitations as applied to foods and grains. In Advances In Drying. Vol 1. Mujumdar, A.S. ed. Hemisphere Publishing Corp., New York, NY. p119-154.

Gariboldi, F. 1974. Rice parboiling. Agr. Dev. Paper No. 97. FAO. Rome, Italy.

Gauvin, W.H. and M.H. Costin. 1980. Spraydrying in superheated steam: a techno-economic study. In Drying 80. Mujumdar, A.S. ed. Hemisphere Publishing Corp., New York, NY. p320-331.

Gauvin, W.H., S. Katta and F.H. Knelman. 1975. Drop trajectory predictions and their importance in the design of spray dryers. Int. J. Multiphase Flow 1(8)793-816.

Gauvin, W.H. and S. Katta. 1976. Basic concepts of spray dryer design. AIChE J. 22(4)713-724.

Gilliland, E.R. and T.K. Sherwood. 1933. The drying of solids. Ind. Eng. Chem. 25(6)1134-1136.

Henry, P.S.A. 1939. Diffusion in absorbing media. Proc. R. Soc. London A1(71)215-241.

Kelly, J.J. and P.O'Donnell. 1977. Residence time model for rotary drums. Trans. Inst. Chem. Engr. 55(3)243-252.

Kerkhof, P.J.A.M. and H.A.C. Thyssen. 1977. Quantitative study of the effects of process variables on aroma retention during the drying of liquid foods. AIChE Symp. Series 73(163)33-45.

Krischer, O. 1938. Grundgesetze der Feuchtigkeitsbewegung in Trocknungsguter: Kapillarwasserbewegung und Wasserdampfdiffusion. VDI Z. 82(2)373-378.

van der Lijn, J. 1976. Simulation of heat and mass transfer in spraydrying., Ph.D thesis. Agricultural University, Wageningen, The Netherlands.

Liou, J.K. 1982. An approximate method for non-linear diffusion applied to inactivation during drying. Ph.D thesis. Agricultural University, Wageningen, The Netherlands.

Lomauro, C.J., A.S. Bakshi and T.P. Labuza. 1985. Evaluation of food moisture sorption isotherm equations. Part 1: fruit, vegetable and meat products. Lebensm. Wiss. u. Technol. 18(2)111-117. Part 2: milk, coffee, tea, nuts, oilseeds, spices, starchy foods. Ibid 18(2)118-124.

Luikov, A.V. 1966. Heat and Mass Transfer in Capillary Porous Bodies. Pergamon Press, New York, NY.

Luyben, K.Ch.A.M., J.J. Olieman and S. Bruin. 1980. Concentration dependent diffusion coefficients derived from experimental drying curves. In Drying '80. Mujumdar, A.S. ed. Hemisphere Publishing Corp., New York, NY. p233-243.

Manning, W.P. and W.H. Gauvin. 1960. Heat and mass transfer to decelerating finely atomized sprays. AIChE J. 6(2)184-190.

Masters, K. 1985. Spraydrying Handbook. 3rd ed. John Wiley and Sons, New York, NY.

Mujumdar, A.S. ed. 1980. Drying '80. Vol 1 and 2. Hemisphere Publishing Corp., New York, NY.

Mujumdar, A.S. ed. 1982. Drying '82. Hemisphere Publishing Corp., New York, NY.

Mujumdar, A.S. ed. 1984. Drying '84. Hemisphere Publishing Corp., New York, NY.

Parry, J.L. 1985. Mathematical modeling and computer simulation of heat and mass transfer in agricultural grain drying: a review. J. Agric. Eng. Res. 32(1)1-29.

Pixton, S.W., R.W. Howe. 1983. The suitability of various linear transformations to represent the sigmoid relationship of humidity and moisture content. A review. J. Stored Prod. Res. 19(1)1-18.

Purcell, J.G. 1979. Practical rotary cascading dryer design. The Chemical Engineer 45(7)497-500.

Ranz, W.E. and W.R. Marshall. 1952. Evaporation from drops. Chem. Eng. Progr. 48(3)141-146.

Reay, D. 1979. Theory in the design of dryers The Chemical Engineer 45(7)501-506.

STANDARDS, SPECIFICATIONS, AND TEST METHODS FOR DRY MILK PRODUCTS

JAN PISECKY
A/S Niro Atomizer, Denmark

ABSTRACT

The manufacture of dried milk products does not consist of evaporation and spray drying only. By applying certain pre-treatments, by selecting and modifying the conditions of the drying process by splitting it into spray drying and fluid bed drying, and finally by incorporating some powder after-treatment procedures, it is possible to influence the product properties to a great extent, so as to produce products which are more suitable for a given use than the conventional products. Consequently, the range of available products and the range of milk powder properties has increased considerably.

This paper deals with milk powder properties which are the basic elements of quality specifications, as well as with the test methods for their determination, and the factors controlling them during manufacture.

The quality of milk powders and other dried milk products is defined by quality specifications. They define the quality requirements in terms of maximum or minimum contents of some components, minimum levels of desired characteristics, and maximum levels of permissible defects. Specifications also define the the laboratory methods for the determination of the individual properties. Few international specifications exist for these most important products of international trade.

An example of an international specification is the EEC specification for skim milk powder as shown in Table 1 (ref. 1).

TABLE 1.

Skim milk powder - EEC standard

Properties

a)	Butterfat	max. 1.00%
b)	Moisture	max. 3.5%
c)	Acidity expressed as:	
	- ml 0.1 N NaOH	max. 3.0%
	- lactic acid	max. 0.15%
d)	Lactate	max. 1.50 mg/100 g
e)	Additives	none
f)	Phosphatase	negative i.e. equal to 4 microgram/g reconstituted milk or less
g)	Solubility index	max. 0.5 ml
h)	Scorched particles	max. 15.0 mg i.e. disc B
i)	Total plate count	max. 40,000/g
k)	B.coli	neg. per 0.1 g
l)	Buttermilk test	neg.
m)	Whey test	neg.
n)	Taste and smell	clean
o)	Appearance	white to light yellow colour, no discoloured particles

Another example of such a specification, which until recently was considered an unofficial international standard, (and still is in some countries) is the American Dry Milk Institute specification. A survey of quality grades according to ADMI is shown in Table 2 (skim milk) and Table 3 (whole milk), (ref. 2).

TABLE 2

Specific grading requirements for nonfat dry milk

	Instant	Spray		Roller	
	Extra	Extra	Stand.	Extra	Stand.
Milkfat, %	1.25	1.25	1.50	1.25	1.50
Moisture, %	4.50	4.00	5.00	4.00	5.00
Titratable acidity, %	0.15	0.15	0.17	0.15	0.17
Bact. estimate, per gm	30,000	50,000	100,000	50,000	100,000
Solubility Index, ml	1.0	1.25[1)]	2.00[2)]	15.00	15.00
Scorched part., mg	15.0	15.0	22.5	22.5	32.5
Dispersibility, %	85.0	-	-	-	-

1) 2.00 for powder designated as "high heat".
2) 2.50 for powder designated as "high heat".

TABLE 3

Specific grading requirements for dry whole milk

	Spray			Roller	
	Premium	Extra	Stand.	Extra	Stand.
Milkfat, %	26.00	26.00	26.00	26.00	26.00
Moisture, %	2.25	2.50	3.00	3.00	4.00
Titratable acidity, %	0.15	0.15	0.17	0.15	0.17
Bact. estimate, per gm	30,000	50,000	100,000	50,000	100,000
Solubility Index, ml	0.50	0.50	1.00	15.00	15.00
Scorched part., mg	7.5	15.0	22.5	22.5	32.5
Copper, p.p.m.	1.5	1.5	3.0	1.5	3.0
Iron, p.p.m.	10.0	10.0	15.0	-	-
Oxygen, % 1)	2.0	3.0	-	-	-

1) Requirements for gas packed products only.

Similarly there is a number of national specifications. Both international and national specifications are characterized by a number of requirements which certify that the products have the proper composition, that they are not affected by undesirable bacteriological or chemical processes, not excessively bacteriologically contaminated, and not falsified by the addition of foreign ingredients.

In addition, there is a great number of manufacturers' and buyers' specifications with many additional criteria of other chemical, physical, and especially functional requirements. Such specifications state not only composition and quality of the product from a health hazard point of view, but also ensure that a given product is suitable for the intended final use. Therefore factories usually have specifications for products, similar in composition but differing from each other in special properties depending on whether they are to be used in households, for further processing in bakeries, chocolate manufacture, for rewetting, or possibly for recombination, etc.

The basic elements of quality specifications are milk powder properties, a survey of which is shown in Fig. 1.

Milk powder properties as shown in the survey are divided into two main groups, i.e. general properties and properties specifically related to milk powders.

FIG. 1: **PROPERTIES OF MILK POWDERS**

General Properties	Properties Specifically Related to Milk Powders		
A) General Requirements	Physical	Functional	Faults
a) chemical b) microbiological c) organoleptic B) Composition of the Solids a) fat b) non-fat solids c) food ingredients d) functional additives	a) free moisture b) particle size distribution ab) primary bb) secondary c) particle density d) bulk density	a) instant properties aa) wettability ba) dispersibility ca) sinkability da) rate of hydration b) flowability c) friability d) WPNI e) hygroscopicity f) " cakyness " g) thermostability h) shelf life	a) solubility index b) scorched particles c) free fat d) scum e) white flecks f) specks

A) General Properties

This includes properties mentioned under International Standards ensuring the required bacteriological quality, freedom from defect, and composition.

B) Properties specifically related to milk powders, i.e. such properties as are directly influenced by the special technology and processes applied in the milk drying industry. These properties can be further divided into:

a) Physical properties which define the structure of the powdered product.

b) Functional properties which define the consumer's requirements, i.e. ensure that the product is suitable for a given purpose.

c) Product faults which indicate some undesirable though often unavoidable deterioration caused by processing.

The distinction between these groups is often not sharp. For example, high free fat is a defect of instant whole milk powder for household use, but is an important functional property of whole milk powder for the chocolate industry. Bulk density, which is listed among the physical properties, is also an important functional property.

In the milk powder industry use of in-line measurement or in-line control is still not common. The process quality control procedure is based on regular sampling and testing of the most important properties by quick laboratory methods, the results of which are used as a feed-back by operators to adjust the processing conditions. The final quality control is done in the factory laboratory where the quality is evaluated according to the appropriate specification.

In order to achieve the desired characteristics and functional properties, it is essential to know the influence of the operating parameters on the individual properties. The time available for this paper does not allow discussion of the whole spectrum of properties. It will therefore deal with some of the properties of the last group (i.e. properties related specifically to milk powders), assuming that the essential factors of final product quality such as raw milk quality, heat treatment, and overall hygienic conditions of the factory are in accordance with the appropriate standard.

Moisture Content

Residual moisture is one of the most important properties of milk powder both from a quality and an economic point of view. The quality specifications lay down the maximum permissible moisture content to achieve the desired shelf-life. The economic aspect demands that maximum moisture content be approached as closely as possible, while at the same time ensuring that no portion of the product will exceed this moisture level.

In two-stage drying, it is also important to control the intermediate moisture, i.e. the moisture of the powder at exit from the drying chamber, because it influences many other properties including solubility index, particle density, bulk density, agglomeration, etc.

It is also important to check occasionally how moisture content changes through the fluid bed system, to ensure that extensive overdrying followed by rehumidificaton to the specified final moisture content in the cooling section is not taking place.

The influence of various factors on powder moisture content ex-chamber is shown in Fig. 2. For some variables the magnitude of influence is known reasonably accurately, whilst for others only the trend is known. The degree of influence is presented as linearily proportional. It should be obvious however that this is not the case, though they may be considered to be such within the relatively narrow operating range under consideration. In regard to the feed total solids content, it is valid up to about 48-50% solids, i.e. in the range where the logarithm of viscosity is linearly proportional to the solids content. Above this solids level, the viscosity increases more rapidly, which in turn influences the residual moisture content.

Because of the importance of powder moisture content, many efforts have been made to control this parameter automatically. Automatic moisture control based on infrared absorption is in operation on many plants, sometimes achieving standard deviation less than 0.1%. Even better results have been achieved when combining the infrared control with feed-forward loops in which the set point of outlet air temperature is adjusted according to variations of the inlet parameters.

The standard laboratory method for determination of powder moisture content is oven-drying at 102 C to constant weight (ref. 3). For process control, a modified quick oven-drying method and laboratory

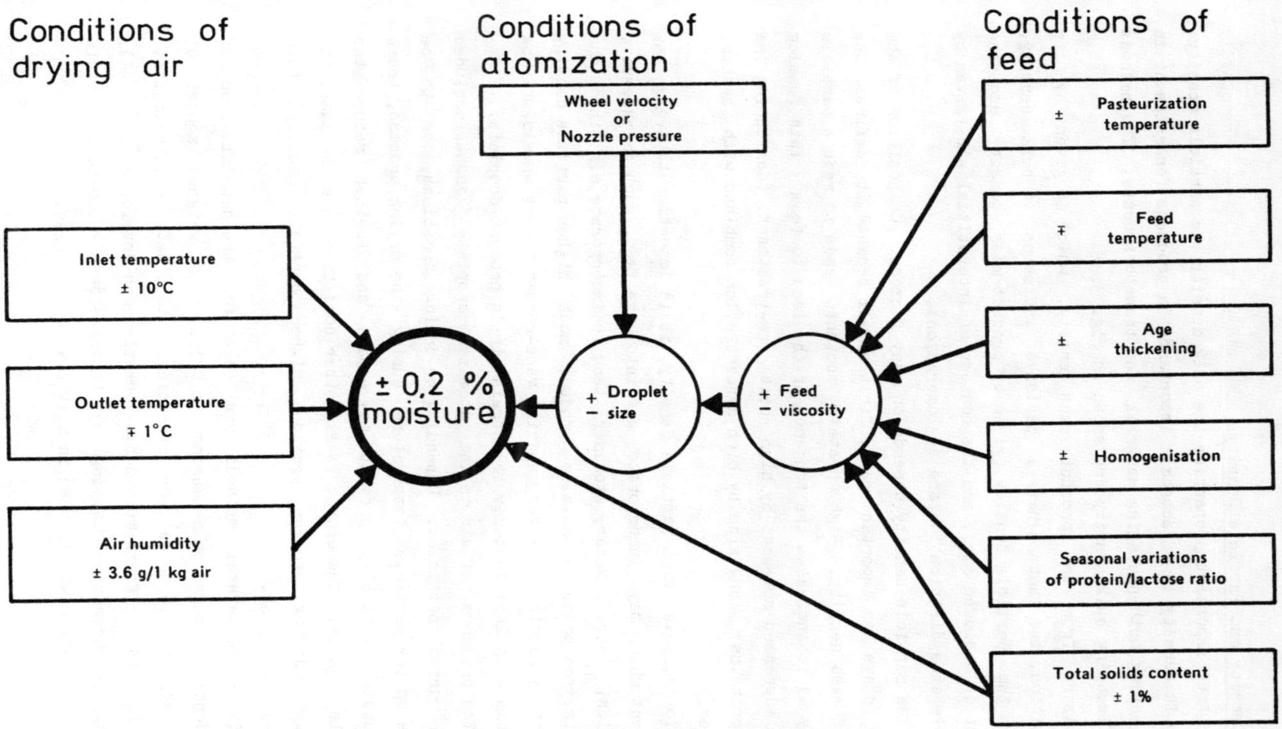

FIG. 2: <u>THE INFLUENCE OF TECHNOLOGICAL VARIABLES ON POWDER MOISTURE</u>

infra-red aparatus are used.

Bulk Density and Particle Density

Other important properties are bulk density and particle density. High bulk density is of economic importance in products for shipment in bulk, saving both packaging material and transport costs. The methods of influencing bulk density are shown in Fig. 3.

Apart from the product mass density, which is given by its composition, two factors have a decisive influence on bulk density, namely the particle density (given by product mass density and the content of occluded air), and the content of interstitial air (given by particle size distribution and agglomeration).

The particle density depends on many factors. Composition of the solids plays an important part, first of all because it defines the product mass density. High protein content tends to reduce particle density as it increases the tendency of the feed to foam. This foaming can be supressed somewhat by high heat pre-treatment, (denaturing the whey proteins), and also by high concentration combined with heating the feed.

To achieve high particle density, it is important to avoid any treatment which may incorporate air into the feed, such as excessive agitation, etc. Rotary atomisers tend to incorporate air into the droplets, and pressure nozzles produce much higher particle density than rotary wheels. However, special vane-shaped rotary wheels are now available with less tendency to entrain air in product droplets.

The presence of air in the atomized droplets causes occluded air in dried particles. Depending on drying conditions, or to be precise on the particle temperature during the drying process, those air bubbles initially present may expand and further reduce the particle density. Therefore, if all other conditions are the same, the two-stage drying process provides higher particle density than single-stage drying.

There are several methods available for the determination of bulk density, based on measuring the volume of a weighed amount of powder subjected to manual or mechanical tapping. An International Dairy Federation (IDF) method has recently been developed (ref. 4), which applies mechanical tapping by the Engelsman volumeter, and this method is becoming widely used in milk powder factories.

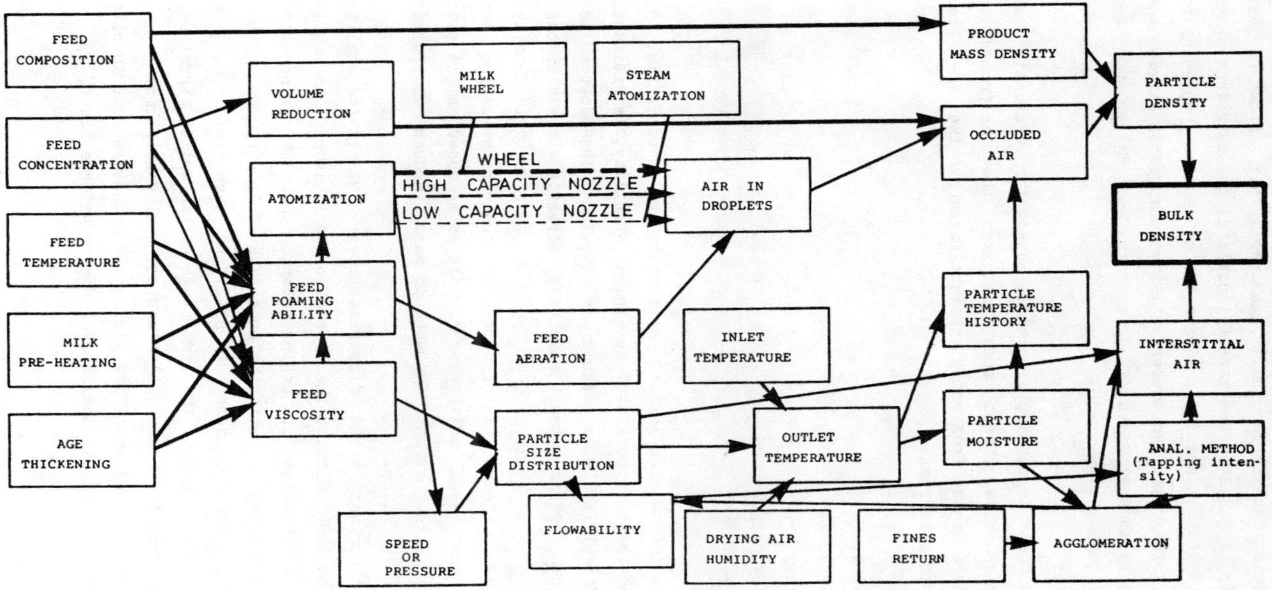

FIG. 3: THE INFLUENCE OF VARIOUS FACTORS ON BULK DENSITY

The particle density is usually measured by the Beckmann air pycnometer. This is based on the same principle as the petroleum-ether method, in which the interstitial air in the powder is replaced by petroleum-ether. The air pycnometer utilizes compressed air for the same purpose and is more accurate.

Instant Properties

In the attempt to develop a suitable method for evaluation of reconstitution properties, it has been recognized that the process of dissolving powders is far from being straight-forward. The so-called "instant" properties include:

- wettability,
- sinkability,
- dispersibility,
- rate of hydration, and
- solubility

The factors influencing instant properties are shown in Fig. 4. Friability, which is the opposite of mechanical stability, is included, because high friability will result in the breaking down of the agglomerates into fines, thus adversely affecting the instant properties.

Wettability is the ability of a heap of powder to be wetted and to penetrate the surface of still water. It is measured as the time necessary for a given amount of powder to pass through the water surface at room temperature.

Even more important is the dispersibility. This is the ability of powder to be wet without the formation dry lumps, and with simultaneous disintegration of agglomerates. The IDF method for dispersibility specifies a stirring procedure following which the solution is poured through a mesh of specified size (ref. 5). The dispersibility is expressed as the percentage of the solids dissolved, and is determined from the total solids content of the solution.

Other methods are used which are similar in principle but involve an examination of the residual undissolved powder using a sieve.

FIG. 4. FACTORS INFLUENCING INSTANT PROPERTIES.

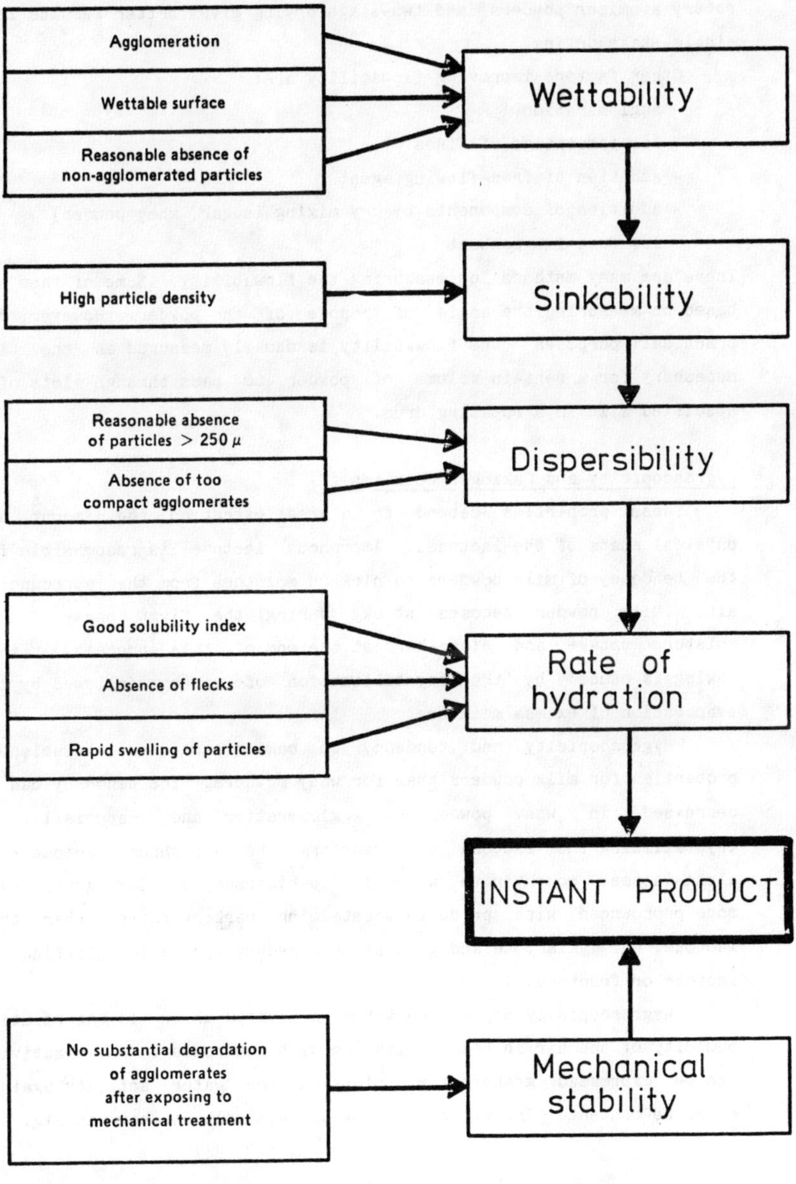

Flowability

This is an important but complex property. The main factors controlling flowability are particle size distribution and surface texture. Pressure nozzle powders are superior in this respect to rotary atomiser powders, and two-stage drying gives better results than single-stage drying.

Other factors improving flowability are:

- agglomeration
- low percentage of fines
- addition of free-flowing agent
- addition of components by dry mixing (sugar, whey powder)
- low free-fat content

There are many methods for measuring the flowability. Some of them are based on measuring the angle of repose of the powder. However, for practical purposes the flowability is usually measured as the time necessary for a certain volume of powder to pass through slots of a specified size in a rotating drum.

Hygroscopicity and Caking Properties

These properties depend to a great extent on the amount and physical state of the lactose. Amorphous lactose is responsible for the tendency of milk powders to pick up moisture from the surrounding air. Milk powder becomes sticky during the first phase of this moisture uptake and stone-hard at the end of this process. Final caking is caused by the crystallization of lactose followed by the evaporation of excess moisture.

Hygroscopicity and tendency to caking are less troublesome properties for milk powders than for whey powders. The tendency can be decreased in whey powder by agglomeration and especially by crystallizing the lactose to transform the amorphous lactose into alpha-lactose monohydrate, which is non-hygroscopic. Caking is even more pronounced with products containing carbohydrates other than lactose, eg. galactose and glucose obtained by enzymatic splitting of lactose or fructose.

Hygroscopicity depends on water activity which is the relative humidity of the air in the equilibrium with the powder. Water activity can be expressed graphically, plotting the water activity against moisture content. The water activity of skim milk is shown in Fig. 5.

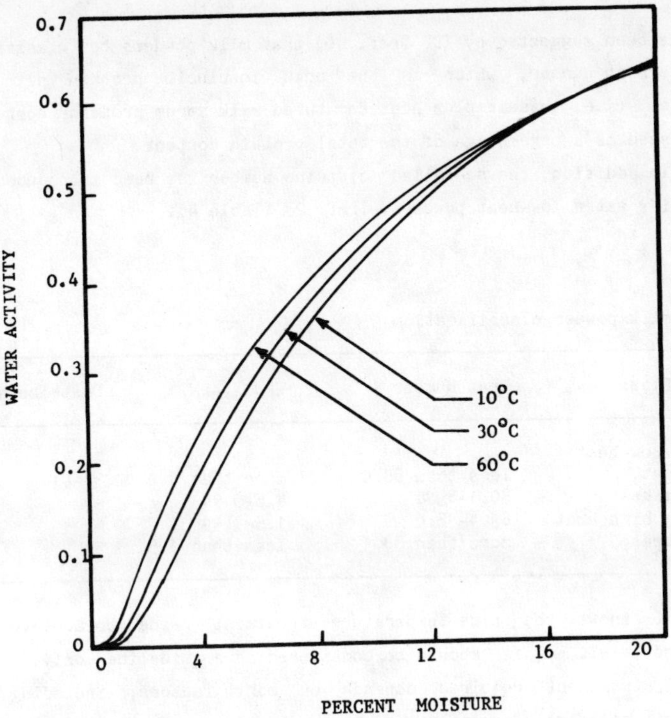

FIGURE 5. Water activity of skim milk as a function of moisture and temperature.

It can be seen that water activity is dependent on temperature.

For practical purposes, hygroscopicity is measured as the equilibrium moisture content of the powder which has been exposed to air of 80% RH. The tendency to caking is then determined and expressed as the percentage of powder which will remain on a defined mesh, after re-drying and sieving the cake formed in a hygroscopicity test.

Whey Protein Nitrogen Index and Heat Number

The classification of skim milk powders according to the American Dry Milk Institute is as follows:

Low heat	More than 6.0 mg WPN/g
Medium heat	1.5-5.99 mg WPN/g
High heat	less than 1.5 mg WPN/g

It has been suggested by IDF (ref. 6) that milk powders be classified by the heat number, which is the acid insoluble protein nitrogen content (i.e. casein plus heat denatured milk serum protein content) expressed as a percentage of the total protein content.

In addition, the so-called cysteine number has been introduced to classify extra low-heat powders, (ref. 2) (Table 4).

TABLE 4.

Skim milk powder classification

Heat Class	Heat Number %	WPNI (ADMI)	Cysteine Number
Extra low-heat	-	-	26-33
Low heat	less than 80.0	more than 6.0	34-41
Medium heat	80.1-83.0	4.5-5.99	-
Medium high heat	83.1-88.0	1.5-4.49	-
High heat	more than 88.1	less than 1.5	-

Fig. 6 shows the time-temperature treatment needed to achieve any desired WPNI. This should be considered as a guideline only. The actual treatment required depends on many factors, including the composition of milk as influenced by the breed, season, climate, etc., and the equipment including type and construction of the heater. The influence of spray drying on WPNI is negligible.

Thermal Stability

Products intended for use as whiteners in hot beverages such as coffee and tea, must be reasonably thermostable. The lack of thermostability occurs as so-called feathering, i.e. particles visible on the surface or throughout the beverage, or as sediment at the bottom.

The means available for controlling thermostability are as follows:

- High-heat treatment of the raw milk
- Addition of stabilizers or adjustment of pH
- Heat treatment of concentrate
- Removal of whey proteins (caseinate-based product)

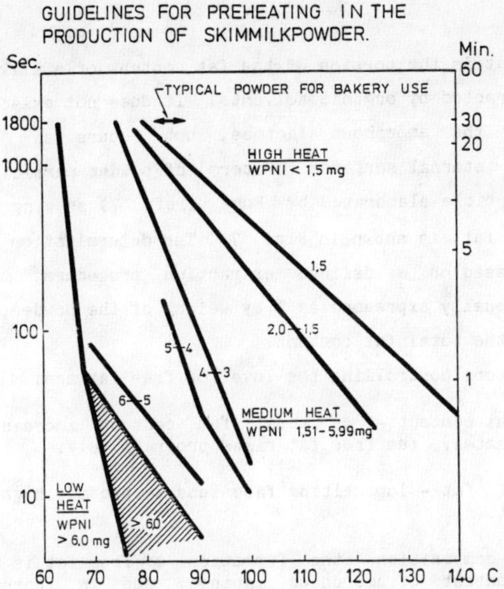

FIG. 6

The test for thermostability is similar to that for solubility index, i.e. by measuring the volume of the sediment created when milk powders are dissolved in a defined hot coffee solution (coffee test), or hot water (hot water test).

Insolubility Index and White Flecks

The term "insolubility index" instead of "solubility index" was introduced when the IDF modification of the original ADMI method was issued. It is determined as the volume of the insoluble sediment after dissolving and centrifuging according to a prescribed procedure.

It is often considered as one of the most important quality criteria. The occurrence of insoluble sediment is undesirable, but to some extent unavoidable. The main factor causing solubility defects is the particle temperature during the water removal process (so called particle temperature history).

The occurence of white flecks is of similar origin to insoluble sediment. Unlike insolubility, this defect can be detected visually but is difficult to determine quantitatively.

Free Fat

Free fat is the portion of the fat content of a milk powder which can be extracted by organic solvents. It does not exist in globules protected by the amorphous lactose, but occurs as "pools" on the external and internal surfaces (craters) of powder particles. A model of powder particle elaborated by Buma (ref. 7) showing four forms of extractable fat is shown in Fig. 7. The determination of free fat content is based on a defined extraction procedure. The free fat content is usually expressed as % by weight of the powder, sometimes as per cent of the total fat content.

The factors controlling the level of free fat are:

a) Total fat content - As total fat content increases above 28% approximately, the free fat rises progressively.

b) Type of fat - low melting fats tend to produce higher levels of free fat.

c) Product composition - the formation of free fat is reduced by a high content of amorphous lactose, and is increased by high protein content.

d) Gentle drying conditions.

e) Gentle powder treatment

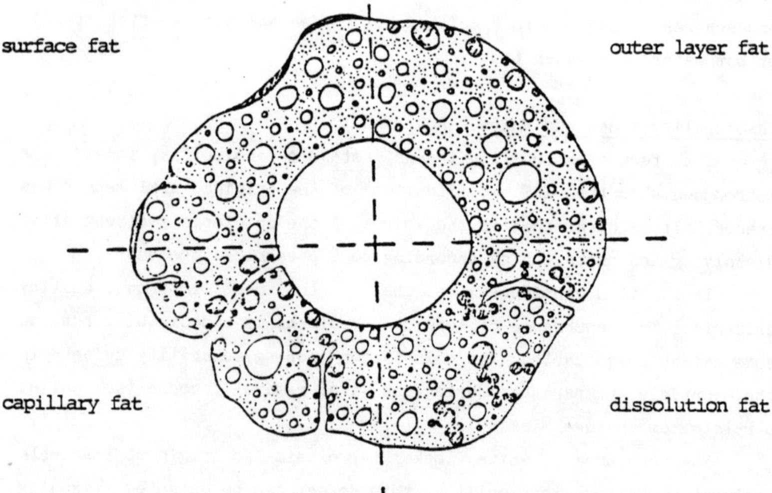

Fig. 7 BUMA's model of 4 forms of extractable fat in a milk powder particle

- avoiding pneumatic transport
- low pressure drop in cyclones
- effective cooling of powder by fluid bed.

f) The powder moisture should not be too low.

g) Avoiding lactose crystallization.

h) Addition of emulsifier.

Table 5 shows a factory quality specification for instant whole milk powder, showing functional quality criteria, but not chemical, bacteriological and organoleptic requirements.

TABLE 5

Example of factory quality specification for instant whole milk powder (functional properties).

Property	Unit	Analytical Method	Target	Reject Level	Ref.
1. Butterfat	%	IDF 9B	28.2	less than 28.0	8
2. Moisture	%	IDF 26	2.7±0.1	more than 3.0	3
3. Acidity (l.a.)	%	ADMI		more than 0.15	2
4. Solubility index	ml	ADMI	less than 0.2	more than 0.5	2
5. Scorched particles	disc	ADMI	A	more than B	2
6. Bulk density	g/ml	NA-A2b	0.48 0.2	outside target	9
7. Free fat	%	NA-A10a	less than 1.5	more than 2.5	9
8. Wettability	s	NA-A5a	less than 15	more than 30	9
9. Dispersibility	%	IDF	more than 95	less than 85	5
10. Particle density	g/ml	NA-A11a	more than 1.18	none	9
11. WPNI	mg/gN	ADMI	less than 1.0	none	2
12. Flowability	s	NA-A23a	less than 40	more than 80	9
13. Coffee test	ml	own	less than 1.0	more than 2.0	-
14. Hot water test	ml	own	less than 2.0	more than 4.0	-
15. White flecks	mm	NA-A20a	less than 1	more than 2	9
16. Phospholipids	%	NA-A22a	0.2±0.02	none	9
17. Particles smaller than 100 microns	%	NA-A8a	less than 15.0	none	9
18. Particles greater than 500 microns	%	NA-A8a	less than 3.0	none	9

An understanding of the basic factors influencing milk powder properties is necessary for the development of new products, as well as for trouble-shooting. Such problems can be very complicated in practice when a range of properties has to be achieved. The technologist is very often faced with a situation where a factor which has a positive influence on one property damages another property or reduces profitability.

In conclusion it is to be emphasized that if milk powder technologists are to deal effectively with these problems, they must have not only a background of technological knowledge, but also adequate flexibility in the production plant.

REFERENCES

1 De Europaeiske Faellesskaber Tidende nr. L 84/19: Kommissionens Forordning nr. 625/78, appendix 1128/84.

2 Standards for Grades of Dry Milk, Bulletin 916 (revised), ADMI Inc., Chicago, 1971.

3 Determination of the Water Content of Dried Milk, FIL-IDF 26: 1964

4 Physical properties of dried milk (determination of bulk density) FIL-IDF E-Doc 204: 1984.

5 Instant dried milk - Determination of the dispersibility and wettability FIL-IDF 87: 1979.

6 Draft International Standard, ISO/DIS 8156.

7 T.J. Buma, Neth. Milk Dairy J., 25(1971) 159-174.

8 Fat in dried milk, dried whey, dried buttermilk, (revision of std. 9A), FIL-IDF 9B: 1984.

9 A/S Niro Atomiser, Analytical Methods for Dry Milk Products, 4th Edition, Denmark 1978.

RECENT DEVELOPMENTS IN SPRAY DRYING OF MILK

M.E. KNIPSCHILDT

APV Anhydro A/S, Denmark

INTRODUCTION

Of all food products, milk is probably the one that is dried in largest quantities. In the countries where milk production is highly organized, the position in recent years has been that more milk is processed than is sold as liquid milk. Sales of liquid milk are slipping in most countries in Western Europe for a variety of reasons: price, changes in life-style, doorstep delivery being abolished etc. Conversion of the milk to powder has become a manufacturing process of equal importance to the dairy industry as the traditional processes, manufacture of butter and cheese. Further, the drying process is used to take care of the by-products from manufacture of these products, namely skim milk and whey. Dried milk manufacture has grown into a large industry of considerable international importance.

In the countries where there is a surplus of milk, the drying process has made it possible to export the milk solids all over the world and make milk proteins available in regions that are unfavourable to dairying or where milk production is insufficient. Recombining dairies have been built in the Far East, in the Middle East and in Africa where the population - thanks to the drying industry - have milk available that has been produced in other parts of the world. The dry milk industry renders a significant contribution to solve the world's food problem.

With modern techniques it is possible to convert the milk to powder without serious loss in nutritive value, i.e. milk made from powder has approximately the same food value as fresh milk.

The sharp rise in energy costs has been a challenge to the milk powder industry and to the manufacturers of evaporators and dryers, because evaporation and drying are the most energy intensive areas of manufacture in the dairy industry. No other process has such a high energy demand per ton of finished product. This is due to the fact that approximately 90% of milk is water, which has to be removed by evaporation by application of heat.

High energy costs have brought about a new generation of evaporators and dryers, making it possible to convert the milk to powder with less than half of the energy used before. The evaporation and drying equipment installed today is very much different from the design used only a few years ago.

CONCENTRATE

Evaporation plants fall outside the scope of this paper, but it must be mentioned that powder properties are to a great extent influenced by the physical and chemical condition of the concentrate. Evaporation and drying should not be regarded as two separate processes, - the conversion of milk into powder is a combined process consisting of evaporation and drying.

In modern plants, 90% of the water in the milk is removed in the evaporator and only 10% in the spray dryer. However, the energy required per kg water evaporation in the spray dryer is much higher than the energy required per kg water evaporation in the evaporator. The fact that only 10% of the water is removed in the spray dryer should not suggest that the efficiency of the spray dryer is of minor importance.

In absolute terms the energy consumption of the dryer is considerably higher than the energy consumption of the evaporator. In a modern plant, only 40% of the total energy consumption is used in the evaporator to remove 90% of the water, and 60% of the total energy requirement is used in the spray dryer to remove only 10% of the water. The evaporator is much more efficient than the spray dryer.

Degree of Concentration

In an effort to improve the overall efficiency of the process, the tendency has been to concentrate the milk as much as possible prior to drying. In modern plants, skim and whole milk are concentrated to 48-50% solids and whey to 55-60%.

It would be advisable to develop evaporators and dryers that can handle milk of higher total solids. However, the use of higher concentrations is not an engineering problem related to the design of evaporators and dryers, - it is a technological problem. Experience shows that if milk is concentrated beyond the above mentioned concentrations, the quality of the powder suffers, - especially the solubility.

The heat stability of the casein is decisive for the solubility of the powder. Poor solubility is usually due to coagulation of the casein. Casein stability is extremely sensitive to variations in pH and salt balance, and both are essentially affected by concentration of the milk. The titratable acidity increases very nearly in direct proportion to the concentration, and also the hydrogen ion concentration is increased. Whereas the pH of fresh milk is approx. 6.7, the pH of 48-50% concentrate is approx. 6.1. Only a very slight increase in the acidity of the milk is enough to decrease the heat stability of the casein, as a change in pH of the milk is magnified by concentration. Although far from the isoelectric point, a pH may be reached where fine casein precipitation starts taking place, which results in poor solubility of the powder. The interaction of such factors as protein content of the fresh milk, concentration, concentrate temperature, acidity, and salt balance determines the degree of casein stability.

There are indications that the concentrations used up to now are not far from the optimal conditions. When these concentrations are used, seasonal variations in the composition of the milk are often sufficient to cause destabilization of the proteins. The stability is restored by decreasing the concentration, - a reduction of only 1-2% in total solids is usually sufficient.

It is a well-known fact that the lactation period affects the composition of the milk. It is also changed when the cows are moved from stable to pasture. The protein content will rise which markedly increases the viscosity of the concentrate, especially when the milk has been exposed to severe heat treatment.

It is desirable to have sufficient flexibility between evaporator and dryer to maintain the capacity when the level of concentration has to be varied over the year. When whole milk has to be dried, it is very good policy to design the evaporator for 50% solids and the spray dryer for 48%. During the time when the composition of the milk permits the maximum concentration to be used, it is easy to reduce the capacity of the dryer by reducing the air inlet temperature.

The chemical and physical properties of concentrate with high total solids greatly affect the properties of the resultant powder. Therefore, before an attempt is made to use higher concentrations, it is necessary to study more closely the factors that affect the properties of concentrate with increasing total solids.

Concentrate Heating

The concentrate is usually discharged from the evaporator at a temperature of about 45°C unless backwards flow or a separate finisher is used in which case the concentrate temperature may be higher.

The viscosity of concentrated milk is significantly affected by temperature (Fig. 1). It is normally desirable to feed the concentrate to the atomizer at the lowest possible viscosity. To achieve this, the concentrate has to be heated to 70-75°C prior to atomization, whereby the following advantages are obtained:

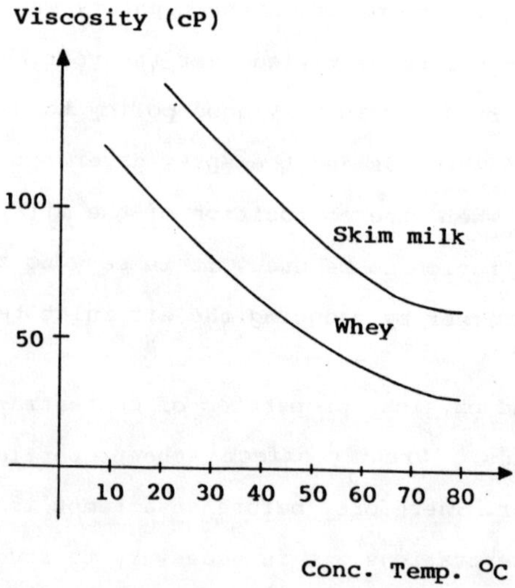

Fig. 1 The effect of temperature on the viscosity of concentrated skim milk and whey.

(1) Saving of energy as it is more efficient to heat the concentrate by means of steam or hot water than to use heat from the air supplied to the dryer.

The evaporation capacity of the dryer is increased by approx. 5% when the concentrate is supplied to the atomizer at 70°C instead of 50°C.

(2) The average size of the droplets into which the concentrate is atomized is reduced due to the lower viscosity. This facilitates the drying making it possible to reduce the air outlet temperature and still maintain the specified moisture content of the powder. The lower outlet temperature results in increased capacity (i.e. lower heat consumption per ton of powder) and more gentle drying which results in improved powder quality, - especially improved solubility. The last point is of particular importance when concentrate with high total solids is dried in a single stage spray dryer, as the outlet temperature can get near the critical temperature limit for solubility.

(3) As the temperature conditions in the last stages of the evaporator and in the feed tank for the dryer can be favourable for growth of microorganisms, heat treatment of the concentrate immediately before drying will contribute to safeguard the bacteriological quality of the powder.

Heating of the concentrate must be done in such a way that no holding takes place, i.e. the concentrate heater must be placed immediately before the atomizer. This is necessary in order to avoid the so-called "age thickening" which occurs very rapidly at temperatures above 60°C. The time-dependent thickening of hot concentrate is irreversible and results in very high viscosity.

If the concentrate heater is placed at the floor in the drier room, the holding time in the feed line to the atomizer can be sufficient to cause thickening of the concentrate. The concentrate heater should be inserted in the feed line between feed pump and atomizer, and it should be placed at the top of the drying chamber next to the atomizer. A tubular heat exchanger of the coil type where the heating medium is vacuum steam or hot water gives satisfactory service when the temperature difference between product and heating medium is small and the product has a high velocity. A scraped surface heat exchanger may also be used, but if the concentrate heater should be suitable for nozzle atomization, a tubular heat exchanger can more easily be designed for the high pressure.

Reverse Osmosis

Instead of evaporating the water it is possible to mechanically remove some of the water from the milk by using membrane techniques. Whereas ultra-filtration is a separation process, reverse osmosis (RO) or hyper-filtration is a concentration process, removing essentially only the water. The fact that the concentration takes place without exposing the product to heat is of no importance in this context, because only low concentration can be achieved and, therefore, RO must be followed by thermal evaporation. For products that are so heat-sensitive that thermal evaporation cannot be used, as for example egg-white, it is essential that the water removal takes place in a RO plant without the use of heat.

Milk and whey can be concentrated by RO in proportion 1:2, i.e. skim milk from 9% solids to 18% and whey from 6% solids to 12%. To illustrate the proportion between thermal evaporation alone and RO plus thermal evaporation we assume as an example that 10,000 kg skim milk at 9% solids should be concentrated to 50%.

Thermal evaporation alone:

10,000 kg	skim milk, 9% solids
8.200 kg	water removed in evaporator
1,800 kg	concentrate, 50% solids

RO plus thermal evaporation:

10,000 kg	skim milk, 9% solids
5,000 kg	water removed by RO
5,000 kg	concentrate, 18% solids
3,200 kg	water removed in evaporator
1,800 kg	concentrate, 50% solids.

Approx. 60% of the total water removal takes place in the RO plant.

The difference in the costs of an evaporator to take the full load and an evaporator to take only 40% of the evaporation will not cover the cost of the RO plant. Therefore, the justification for RO must be a saving in the running costs.

The driving force in RO is electric power for the circulation pump. The justification for removal of 60% of the water by RO would be that the cost of the power required by the RO plant would be less than the cost of the steam required by a TVR evaporator to remove the same amount of water.

However, in the cases where electricity should be used for concentration rather than steam, it is more advantageous to install a MVR evaporator than an RO plant and do the whole evaporation in one machine. The power consumption of a modern MVR evaporator per kg water evaporation is about the same as for an RO plant and, therefore, MVR is the answer rather than RO when electric power should be used for concentration of milk.

With the development of MVR evaporators, the interest in using RO in the dairy industry has decreased.

The running costs of an RO plant is not only the power consumption. Maintenance in terms of replacement of membranes is a significant cost and the cost of enzymes for cleaning of cellulose acetate membranes should also be included.

Another problem, which seldom is appreciated at the time an RO plant is purchased, is that the loss of milk solids into the water penetrating the membranes (the permeate) is at a level where disposal of the permeate presents a problem. Experience has shown that the biological oxygen demand (BOD) of the permeate is too high to permit disposal of the permeate into rivers and lakes. It must be discharged into an effluent water disposal plant whereby costs are incurred. The condensate from an evaporator will normally be sufficiently pure to allow direct disposal, but normally the condensate is used for cleaning instead of hot water. However, the permeate from a RO plant is cold and has, therefore, no value for cleaning purposes.

In special cases, if for example whey from cheese or casein plants has to be transported to a drying station in a different location, it may be feasible to use RO at the location where the whey is produced and thereby reduce the volume to half of the original volume, but normally RO will not be able to compete with thermal evaporation in the dairy industry.

Thermal Efficiency

The thermal efficiency of a spray dryer is the ratio between the heat utilized for evaporation and the heat supplied to the drying air. The heat required for secondary drying is normally not included in this context, and therefore, the thermal efficiency in percent is expressed by

$$\frac{Ti - Tu}{Ti - Ta} \times 100 \quad \text{where}$$

Ti is air inlet temperature

Tu is air outlet temperature

Ta is ambient air temperature

It is apparent that the efficiency is improved with raising Ti and Ta and with falling Tu.

The max. air inlet temperature to be used in a milk dryer is 215-220°C.

The ambient air temperature is raised by employing heat recovery.

The aim to increase the efficiency by reducing the outlet temperature has led to the introduction of two- and three-stage drying. Multistage drying is based on the fact that whereas most of the water is readily evaporated from the concentrate, a more severe heat treatment, - higher temperature or longer drying time, - is required to remove the last few percentages of the moisture. The rate of evaporation is quick until a dry particle surface is reached. It takes more heat or longer time to evaporate the moisture contained inside the particles.

Two- and Three-stage Drying

Fig. 2 shows a two-stage dryer with fluid bed secondary drying. The powder is discharged from the drying chamber at 5-6% moisture into the vibrating fluid bed dryer and cooler. The fluid bed consists of a stainless steel perforated plate through which air is blown upwards. The powder is "fluidized" by the air, and high turbulence occurs which provides excellent conditions for heat transfer between air and powder.

In the first section, the powder is dried to its final moisture content by air of 100-120°C, and in the second section the powder is cooled by air of approx. 10°C. The powder layer is about 100mm high and the fairly long residence time in the fluid bed (10-12min.) makes it possible to perform the drying at low air temperature. The air used for drying and cooling is collected in the hood over the perforated plate from whence it goes to the cyclones for separation of the fines carried with the air.

Fig. 2 Two stage dryer with solids recovery and heat recovery.

1 Textile filter
2 Heat exchanger for heat recovery
3 Booster in flue gas stack
4 Air preheater
5 Fluid lift system to powder silo

(By courtesy of APV ANHYDRO A/S, Denmark.)

The high moisture in the powder from the chamber is obtained by using a low air outlet temperature. The low air outlet temperature improves the thermal efficiency of the dryer, and it also improves the powder quality, - specially in terms of solubility, bulk density, and free fat.

The dryer in Fig.2 is equipped with solids recovery (textile filter) and heat recovery. Liquid is used to transfer the heat from the outgoing air to the intake air, and the temperature of the liquid is boosted in the flue gas stack for the indirect oilfired air heater.

Three stage drying (Fig. 3) was introduced in recent years in order to further improve the thermal efficiency of the drying process by transferring a greater portion of the

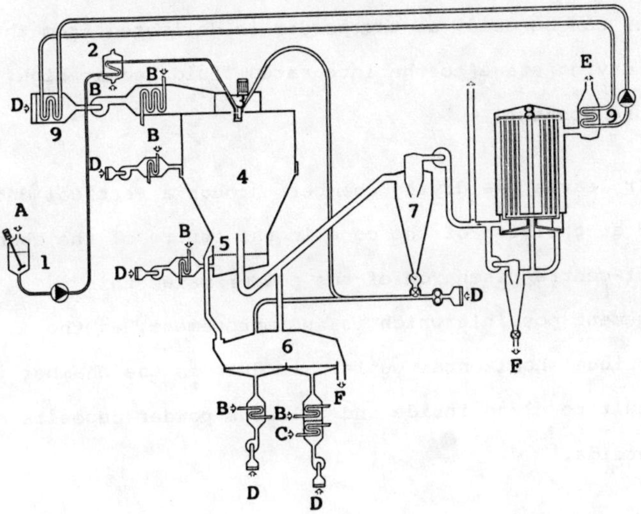

Fig. 3 Three-stage dryer with solids recovery and heat recovery.

1 Feed tank
2 Preheater
3 Atomizer
4 Spray drying chamber
5 Integrated fluid bed
6 External fluid bed
7 Cyclone
8 Bag filter
9 Liquid coupled heat exchanger

A Product inlet
B Steam
C Cooling water
D Air inlet
E Air outlet
F Product outlet

By courtesy of APV ANHYDRO A/S, Denmark.

evaporation from the first stage to the second and third stage. The second stage is a fluid bed placed at the outlet of the conical part of the drying chamber. This fluid bed is static, (non-vibrating), and the semi-dry powder drops over an adjustable weir into an external fluid bed for final drying and cooling. A low outlet temperature from the drying chamber can be used, as the powder is delivered from the first drying stage to the integrated fluid bed at high moisture content.

The air leaves the drying chamber through a vertical duct placed at the base of the cone in the centre of the chamber. The off-centre discharge of the powder makes this arrangement possible which is an improvement, as the conventional horizontal outlet air duct in the chamber is difficult to clean inside and collects powder deposits on the outside.

Further, the horizontal air duct prevents the swirl pattern of the drying air being extended into the cone, as the duct obstructs the rotating movement of the air. Without the horizontal air duct it is possible to support the swirl movement by the so-called wall sweep arrangement, i.e. tangential introduction of secondary air through a number of narrow vertical slits placed in the drying chamber wall at the transition point between the cylinder and the cone. The temperature of the sweeping air is slightly lower than the temperature of the air in the chamber. The high air rotation prevents the moist powder from lodging in the cone, and it also reduces the amount of powder carried with the air to the cyclones. The combination of wall sweep and a steep cone reduce the requirement for external hammers, i.e. the number of hammers as well as their impact can be reduced.

The external fluid bed can be substituted by a pneumatic system for drying and cooling of the powder, if the height under the chamber is insufficient for installation of a fluid bed. This may be the case when an existing single stage dryer is converted to two or three stage drying.

Both the two stage dryer as well as the three stage dryer can operate with either rotating atomizer or pressure nozzle atomizer.

Table 1 shows the heat required to evaporate 1 kg water in single stage, two- and three stage dryers. The air inlet

TABLE 1

Heat Required to Evaporate 1 kg Water.

Air Inlet Temperature 215°C

Type of Dryer	**Kcal**
Single stage	1200
Two stage	980
Three stage	850

temperature in all three cases is 215°C, and the dryers have no heat recovery. As heat recovery reduces the heat consumption by approx. 20%, it will be seen that the heat consumption of a three-stage dryer with heat recovery approaches half of the heat consumption of a single stage dryer without heat recovery. This clearly illustrates the progress made in energy conservation, and the saving per ton of powder will be even greater when the contribution from a low energy evaporator is included.

The advantages offered by three-stage drying are reduced heat consumption, reduced space demand, and improved powder quality. The amount of air required to achieve a certain capacity is the smallest possible when three-stage drying is used, and hence the cost of a textile filter for cleaning of outgoing air is reduced.

Heat Recovery

Several methods have in recent years been introduced to transfer heat from the outlet air to the inlet air, - either by air to air heat exchangers or by air to liquid heat exchangers.

Air to air heat exchangers can be of the plate type or tubular heat exchangers. The air intake is usually far from the point where the air is discharged, and therefore, an air to air heat exchanger requires considerable ductwork. Fouling of the surface in contact with the outlet air will occur if the heat exchanger is used without a filter after the cyclones. The fouling will reduce the heat transfer, and therefore, the heat exchanger must be cleaned frequently. The fouling does, of course, reduce the amount of powder discharged to the atmosphere, but the milk solids are lost and the frequent wet cleaning adds to the load on the effluent plant. In case of a blocked cyclone, the heat exchanger will be clogged up with powder.

A much simpler heat recovery system can be used when the air from the cyclones is cleaned in a filter. In this case exhaust air is passed over the surface of a finned tube heat exchanger transferring heat to the liquid inside the tubes. The heated liquid, which can be water with glycol, is pumped to an air heater battery in the inlet air stream to preheat the inlet air. It does, of course, involve heat transfer twice, and therefore, the resultant preheat temperature is slightly lower than when an air to air heat exchanger is used, but the system is very much simpler and much less costly. If an indirect oil - or gasfired air heater is used, the temperature of the water in the heat recovery system can be raised by running the water through an economizer in the flue gas stack for the air heater. In this way the water temperature is boosted and a higher preheat temperature is achieved.

It is, of course, most rewarding to recover heat from the outgoing air in a system where the temperature of the outgoing air is high. That is the case when single stage drying is used. If we assume an air inlet temperature of

215°C, ambient temperature 15°C, and a preheat temperature of 70°C obtained by heating the inlet air with the outlet air, it will be seen that instead of heating the air 200°C, the air will only have to be heated 145°C with heat recovery, i.e. a saving of 27,5% is obtained. In a three stage dryer where the temperature of the outgoing air is lower, the saving by heat recovery will be about 20%. The pay-back time for a heat recovery system will be short when the annual utilization of the dryer is high.

Air Heating System

The thermal efficiency referred to earlier does not indicate the overall thermal efficiency of the spray drying process, because the heat loss incurred in heating the drying air is not included. The thermal efficiency of the air heating unit is assessed separately.

A steam air heater has practically 100% efficiency, but to compare with other systems, the efficiency of the boiler must be included. In a modern dryer the air inlet temperature is 215-220°C, and steam is seldom available at the pressure required to produce such high air temperature. When steam heating is used, the air temperature is usually boosted by another air heating system.

A development in recent years has been an increased use of liquid phase heating (hot oil system) to boost the temperature from a steam air heater, or to do all the heating. A high thermal efficiency is obtained by preheating the return fluid from the air heating unit in a heat exchanger placed in the stack for the unit employed for heating the thermal fluid. By inserting an economiser in the stack, the flue gases are cooled to 180-200°C by means of the return fluid at approx. 160°C from the air heater. In this way a system thermal efficiency of close to 90% is achieved, whereas the efficiency of an indirect oilfired air heater is about 85%.

Direct Gas Firing

The availability of natural gas in many countries and the high price of oil have in recent years created considerable interest in this system. Fig. 4 shows a direct gas-fired air heater. No heat exchanger is required. The mixture of

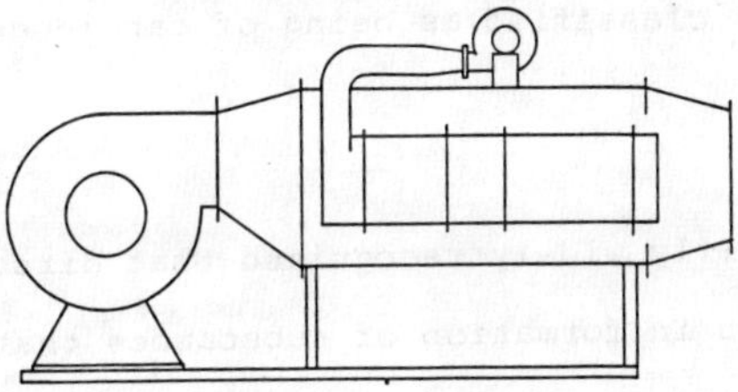

Fig. 4 Direct Gas-fired Air Heater.

the combustion gases from the burner and ambient, filtered air is led to the drying chamber as drying medium. The absence of a heat exchanger brings the thermal efficiency to practically 100%, i.e. a saving in fuel consumption of 10-15% is obtained compared with indirect heating. Further, the capital investment in a direct system is substantially less than for an indirect system. Maintenance costs are lower and the quick response and flexibility of a direct system is also an advantage.

In spite of the attraction offered by direct gas firing, it is not yet widely used outside North America. The reason is recognition in recent years of the danger of contaminating the milk powder by carcinogens.

Research work has shown that an increased content of nitrite and nitrate is found in milk powder prepared in direct gas fired dryers, as much as double that found when indirect heating is used. When amines associated with the proteins react with nitrite, nitrosamines can be formed. This process develops further if nitrate is bacteriologically converted to nitrite. It is a well known fact that nitrosamines are classified as being of carcinogenic nature.

Whereas it is fairly widely recognized that direct gas firing may result in formation of substances that constitute a health hazard, there is widespread ignorance of the reason for it. It has been suggested that it might be due to incomplete combustion or due to impurities in natural gas. The statement has been made that it should be safe to use methane, propane or butane as these gases contain no impurities. All these suggestions represent fundamental misunderstandings. Complete combustion of gas is easy to achieve, and possible impurities in natural gas are stripped off before the gas is supplied to the consumers.

The contamination is produced by the oxides of nitrogen contained in the combustion gases. When nitrogen oxides get in contact with food products it can react with the amines whereby nitrosamines are produced, and nitrogen oxides can also cause formation of nitrite and nitrate. It is not incomplete combustion or impurities in the gas that give cause for concern, - it is the nitrogen oxides formed by the combustion. The key to the problem is to reduce the amount of combustion-generated nitrogen oxides as much as possible. Nitrogen oxides are produced as part of all combustion processes in which air is used as the oxidant. When nitrogen and oxygen are heated, nitrogen oxides are formed by thermal fixation of the nitrogen, - mainly NO and NO_2, but also higher oxides are formed. A common designation for all the nitrogen oxides in the combustion gases is the term NO_x. Conversion of possible chemically bound nitrogen in the fuel will also result in nitrogen oxides in the combustion gases. Fuel bound nitrogen is present in oil and coal but not in gas. When gas burns, all of the NO_x in the combustion gases is generated by the combustion. This is a thermal process which takes place regardless of the kind of gas used.

Food products are in contact with the combustion gases from gas in bakeries, in the gas ovens used in households, and in open grills with the combustion gases from coal. However, as the surface of such food products is small compared with the volume, the formation of nitrite, nitrate and nitrosamines is so small that it is not possible to measure.

It is a different matter when milk is spray dried, because the product is given a very large contact surface in the spray dryer. It has been established that when direct gas fired spray dryers are used, nitrite, nitrate, and nitrosamines are formed in the powder at a rate in direct proportion to the amount of nitrogen oxides in the drying air. Therefore, the answer is to minimize the quantity of nitrogen oxides in the combustion gases. The recognition of this requirement has only taken place in recent years, and it has led to the development of a new generation of gas burners for the food industry, - the so-called low NO_x burner.

The rate of thermal formation of NO_x from oxygen and nitrogen is strongly temperature dependent. The amount of combustion generated NO_x is governed by the flame temperature and the duration of the max. temperature phase. The flame temperature should be as low as possible, but high enough to ensure complete combustion. The max. temperature phase should be as short as possible, but combustion must not be inhibited too early by dilution with the air to be heated, as this will produce undesirable intermediate combustion products.

The flame temperature should not exceed 1200-1300°C and this is achieved by using over-stoichiometric combustion with an excess air level of up to 90%. Uniform mixture of the gas and air is essential to avoid pockets of gas rich mixtures. In the low NO_x burner, the NO_x level ex-burner is guaranteed to be below 1 ppm by volume. Conventional gas burners will normally produce a NO_x level ex-burner of about 150 ppm. The NO_x level of the air getting in contact with the atomized product will be substantially lower than the NO_x level ex-burner due to dilution by the air to be heated. At a NO_x level of 1 ppm ex-burner, the NO_x level of the drying air will be 0,12 ppm at a drying air temperature of 200°C. The figure 0,12 ppm should be understood as increase in NO_x level (NO_x added from the combustion), as ambient air can contain variable concentrations of NO_x produced by industry, power stations and motor cars.

Direct gas firing had been used in milk dryers in North America long before the problems associated with this system were recognized. Skim milk and whey are the main products, and it may be that the problems are more pronounced when powder containing fat is produced (whole milk powder and fat-filled powder), but this is only an assumption which must be verified by further research work.

The food control authorities realize that the use of direct gas firing for manufacture of edible milk powder presents a complex problem, and are, therefore, usually reluctant to make any commitments. It is not allowed, and it is not forbidden. The answer from health authorities on an application to use direct gas firing has been that the heating system used is not subject to any legislation, but it is the manufacturers' responsibility to ensure that the permitted limits for the content of nitrite, nitrate, and nitrosamines in the finished product are not exceeded. Several milk dryers in HOlland are direct gas fired. They are equipped with low NO_x burners and the authorities have the right to demand analysis for nitrite, nitrate, and nitrosamines of samples taken at any time. The content of nitrosamines must be below 1 ppb.

Another product formed by the combustion of gas is water. This does, of course, not constitute a health hazard, but it reduces the capacity of the dryer. The water of combustion must be taken into consideration in the design of the dryer. The economic advantage from direct gas firing is to some extent offset by the loss in capacity caused by the water of combustion.

PROTECTION OF ENVIRONMENTS

A manufacturing creamery can have an undesirable influence on the environment, and the problems have been magnified in recent years by the increase in size of dairy factories. Steps required to protect the environment must be given serious attention. The most important areas are: effluent, noise, and dust. Whereas the ways to control effluent and noise are fairly well known, the problem of dust emission has been neglected for far too long. This problem will be discussed in the following.

Cyclones are normally used to separate the powder from the air in spray dryers. The cyclones are unable to recover the fine powder fraction consisting of particles below 5-10 microns. This powder fraction is usually discharged to the atmosphere with the air. The powder loss from the cyclones is often expressed as percent of the total powder production, but to assess the loss with regard to the environment it is necessary to express it as the amount of solids in relation to the volume of the discharged air, - usually in mg solids per standard cubic metre air.

In most countries legislation has been introduced in recent years to limit the emission levels by specifying the maximum permitted amount of solids contained in the discharged air, - even in the cases where the dust does not constitute a health hazard. In Europe the permitted emission levels from milk driers vary between 50 and 100 mg/st.m^3 air discharged, and in U.S.A. from 25 in California to 90 mg/st.m^3 on the East Coast.

In practice cyclones losses vary within wide ranges depending on several technical and technological factors. Table 2 shows typical cyclone losses from three dairy products.

It will be seen that the loss from cyclone separators in all cases exceeds the permitted emission level. Hence it is not permissible now to install a milk dryer where the powder recovery system consists of cyclones alone. Further cleaning of the air is necessary to bring the loss to the atmosphere within the permitted limit. The powder discharged to the atmosphere represents a significant financial loss, and therefore, when the air is cleaned further it should be done in such a way that the solids recovered can be added to the powder from the dryer to increase the yield.

There are two approaches: wet scrubbing and dry filtration.

Wet Scrubbing

A wet scrubber after the cyclones will trap the greater portion of the fines lost from the cyclones. If a cyclone loss of 200 mg/st.m^3 has to be reduced to 50 mg/st.m^3, the

TABLE 2

Powder Loss From Cyclones

Product	Loss: Percent of Total Production	Loss: mg/st.m^3
Skim Milk	0,5	200
Whole Milk	0,35	150
Whey	1,0	400

wet scrubber must have an efficiency of 75%. This can be achieved by most scrubbers. If the scrubbing medium is water, the milk solids are lost in the water, and although it can be recirculated for some time, the disposal of the scrubber water presents a problem. The scrubber is hygienically unacceptable in a dairy, because it provides ideal conditions for rapid growth of micro-organisms.

To overcome the problems of recovery of the solids and disposal of the scrubbing medium, the so-called milk scrubber using unconcentrated milk as scrubbing medium was reintroduced ten years ago. The milk goes from the scrubber to the evaporator, and the solids recovered will be contained in the milk to be dried and will thus increase the yield.

The milk scrubber originates from U.S.A. where it was abandoned 50 years ago due to the bacteriological hazard. It was reintroduced in Europe about ten years ago by reputable companies claiming that precautions were taken in the new design to prevent bacteria build-up. A further claim was energy saving by preheating of milk and evaporation of a small portion of the water in the milk in the scrubber by means of the heat in the discharged air.

Milk scrubbers were installed on dryers in several countries, but were a serious failure. Today they are practically all removed or used only where powder for animal food is produced. A duplex or triplex installation is required for edible products, as the scrubber must be cleaned after 3-4 hours operation.

There is a wet zone in a scrubber, and at the air exit there is a dry zone. The transition between wet and dry represents a moist zone where the conditions for rapid growth of bacteria are ideal. The number of bacteria in the milk is increased immensely after only a few hours operation. A further risk is the formation of toxins, which can develop when a great number of bacteria are present. Toxins from pathogens like staphylococci are highly toxic, they are not destroyed by the heat treatment commonly used in milk processing, and it is difficult to analyse the powder for toxins.

The use of milk scrubbers represents a considerable bacteriological hazard. It is an unhygienic piece of equipment, which should not be permitted in a dairy. It contaminates the milk and its value as energy saver is negligible. The heat contained in the discharged air should be used to improve the efficiency of the dryer which is the big heat user, and not the evaporator, which is highly efficient. It can increase the yield, but the price is that it places the whole production in jeopardy. The employment of a milk scrubber goes against the common trend towards raising product quality.

Dry Filtration

In this system the milk solids contained in the air from the cyclones are collected in a textile filter (fabric filter or bag collector) before the air is discharged to the atmosphere. The filter is not employed as main collector as is the case in some older dryers, - the cyclones are retained.

A filter meets even the most strict emission levels as the loss is normally guaranteed to be less than 10 mg/st.m^3. If the air from the cyclones contain 200 mg/st.m^3, the filter will recover 95%.

By installing a filter specially designed for milk dryers, a first grade product is recovered, which is continuously blended into the main stream of powder from the dryer and thus increasing the yield. The financial contribution is considerable as the following example shows. We assume a dryer produces 4 ton/h skim milk powder and is in operation 4000 hours per year. With a stack loss of 0,5% and a filter with efficiency 95%, the powder recovered from the filter amounts to 76 ton per year. At a value of £ Sterling 900 per ton, it represents a financial contribution of £ Sterling 68,400 per year.

If the dryer is used for different products with frequent change of product, it is hardly feasible to blend the filter powder into the main production, because it is a major operation to clean the filter, but if the dryer is used only for skim- and whole milk powder it is not necessary to clean the filter when the product is changed, as the automatic bag cleaning system will be sufficient.

The powder from the filter consists of very small particles, i.e. the texture is different from that of the powder from the dryer, and therefore, continuous back-mixing is necessary. Experience shows that the filter powder does not differ from the powder from the dryer with regard to solubility and bacteria count. Sediment tests made on filter powder sometimes show disc B indicating that the filter needs cleaning, but that is without practical importance, because the quantity is too small to influence the sediment of the powder resulting from back-mixing.

The filter must be by-passed when the dryer is wet cleaned, and when the dryer is idle, the filter bags must be protected against moisture pickup, either by supplying hot air to the filter or by circulating dehumidified air through the filter.

Besides providing environmental protection and increased yield, the textile filter serves another important purpose. It provides clean air for heat recovery i.e. transfer of the heat from the outgoing air from the dryer to the inlet air. When the outgoing air contains practically no milk solids, a simple heat exchanger can be installed, as the risk of fouling the surface in contact with the hot air has been eliminated. The requirements of a secondary collector are environment protection, increase of yield and potential for heat recovery, and the filter meets all three requirements.

It should be mentioned that in plants where filter powder is mixed into the powder from the dryer, special attention must be paid to the filter for the inlet air. If the filter for the inlet air is less efficient than the textile filter for the exhaust air, there will obviously be an accumulation of impurities in the textile filter. The filter for the inlet air must be more efficient or at least as efficient as the textile filter for the discharged air. Poor sediment values in the filter powder can be due to an inefficient air inlet filter.

POWDER PROPERTIES

The properties of the powder are affected by the milk quality, the design of the evaporator and dryer, and by the process conditions. Some properties are governed by conditions outside the evaporator and dryer. This is the case when whey is dried,as the desired non-hygroscopic nature of the powder is chiefly governed by crystallization of the lactose between evaporator and dryer. Heat classification of the powder is determined outside the evaporator and dryer, as it is a function of the heat treatment prior to evaporation.

In this section, the influence of drier design and process conditions on powder properties is considered.

Influence of Dryer Design

It is mainly the atomization that influences powder properties. Two systems are used: rotating atomizers and pressure nozzles. The centrifugal force available in a rotating atomizer to break up the liquid is much higher than the force from the pressure in a nozzle atomizer and, therefore, the rotating atomizer can handle more viscous products (i.e. concentrate with higher total solids) than a nozzle atomizer.

Nozzle powder has less occluded air than powder from a rotating atomizer. Lower amounts of occluded air result in higher bulk density of the powder, and reduced oxidation of high-fat powders.

Whole milk powder and fat-filled powder produced by nozzle atomization are claimed to contain less free fat than powder produced by means of rotating atomizer. The high pressure pump and the passage through the nozzle have a homogenizing effect, which will reduce the amount of free fat in the powder. A similar effect can be obtained by homogenizing the concentrate prior to feeding it to a rotating atomizer.

A rotating atomizer can handle concentrate of high total solids and can, therefore, improve thermal efficiency. Nozzle atomizers are not suitable for whey concentrate containing lactose crystals, as the crystals will wear the nozzles. If powder from different products and powder with certain characteristics is to be produced, the dryer must have both atomization systems.

The design of the air distributor and the drying chamber

dimensions influence possible formation of powder deposits and thereby scorched particles in the powder.

The chamber volume and the pattern of the air flow determine the residence time and hence the heat treatment to which the powder is exposed in relation to the chamber temperature. The heat treatment is crucial for the solubility of the powder. Two- and three stage drying result in low chamber temperature and consequently improved solubility.

The influence of direct gas firing on the powder has already been explained.

Feed back of fines to the atomization zone, combined with fluid bed after-drying and cooling, produce an agglomerated powder. Pneumatic after-drying and cooling tend to break down the agglomerates and result in powder with a finer texture.

The tall form dryer, introduced by Morinaga in Japan, is a single nozzle dryer, and it is claimed to give powder of coarse texture.

The filtermat dryer, designed by the Pillsbury Company in U.S.A., produces the dried product as a compact cake which has to be milled, so that a coarse powder is obtained with texture different to that obtained in a conventional dryer.

Influence of Process Conditions

Experiments have been made at the Dairy Research Institute at Melle in Belgium to assess the effect of variations in process conditions on powder properties. The results are quoted below (refs. 1, 2).

In the first experiment (table 3) the influence of raising inlet temperature of the drying air was studied. The inlet temperature was raised from 170°C to 225°C, whereas the outlet temperature was kept constant at 90°C by gradually increasing the feed rate.

The results show that the solubility is only affected very little by raising air inlet temperature when the outlet temperature is kept constant.

The moisture content increases because the air outlet temperature was kept constant. When the air inlet temperature is increased, the outlet temperature will also have to be increased if the moisture content is to remain constant. As a rule, the outlet temperature must be increased by 1°C for each 5°C increase in inlet temperature.

The free fat decreases with increasing inlet temperature, - probably because a hard particle surface is formed as a result of the high drying temperature. The hard surface prevents the fat solvent from penetrating into the

TABLE 3

Influence of Raising Inlet Temperature at Constant Outlet Temperature 90°C. Homogenized Whole Milk Concentrate, 48% Solids.

inlet temp. of the air °C	moisture content %	ADMI-solubility index	Free fat content % of total fat	bulk density g/cm^3
170	2.58	0.05	6.62	0.68
180	2.69	0.05	5.55	0.66
187	2.92	0.05	4.72	0.66
195	2.99	0.10	4.14	0.65
205	3.05	0.15	4.10	0.63
215	3.19	0.30	4.00	0.63
225	3.30	0.30	4.05	0.62

particles so that less fat is extracted when the free fat is determined.

The density decreases with increasing inlet temperature in spite of increased moisture content, - probably because of increased entrapped air in the powder particles.

In the second experiment (table 4) the effect of increasing air outlet temperature was studied. The inlet temperature was kept constant at 195°C, whereas the outlet temperature

TABLE 4

Influence of Raising Outlet Temperature at Constant Inlet Temperature 195°C. Homogenized Whole Milk Concentrate, 48% Solids

outlet temp. of the air °C	moisture content %	ADMI-solubility index	Free fat content % of total fat	bulk density g/cm^3
75	4.75	0.05	2.15	0.64
80	4.41	0.05	2.64	0.63
85	3.93	0.05	3.04	0.61
90	3.23	0.50	2.78	0.60
95	2.59	0.50	4.49	0.58
100	2.18	2.90	4.38	0.57
105	1.76	3.30	5.43	0.55

was raised from 75°C to 105°C by gradually reducing the feed rate.

The moisture content is, of course, reduced with increasing outlet temperature.

The solubility is significantly affected by high outlet temperature. At outlet temperature 100°C and above, the solubility is very poor.

The free fat increases with increasing outlet temperature, which may be ascribed to crack formation in the surface of the particles due to overheating.

It is commonly known that the density falls with falling moisture content.

The low outlet temperatures (75, 80, and 85°C) correspond to the outlet temperature range when two or three stage drying is used. The results clearly illustrate the beneficial effect of multistage drying on solubility and free fat.

The third experiment (table 5) was done to illustrate the

TABLE 5

Influence of Rotation Speed of the Atomizer at Constant Inlet Temperature 195°C and Constant Outlet Temperature 90°C. Homogenized Whole Milk Concentrate, 48% Solids

revolutions of the atomizer per min.	peripheral speed of atomizer disc m/sec.	moisture content %	ADMI solubility index	free-fat content % of total fat	bulk density g/cm^3
19,600	106	3.27	0.90	7.52	0.62
22,250	120	2.82	0.74	5.06	0.59
25,000	135	2.65	0.05	5.57	0.59
28,300	153	2.55	<0.05	6.48	0.58
31,300	169	1.99	<0.05	4.22	0.58

influence of the degree of atomization at constant air inlet and outlet temperatures. The rotation speed of the atomizer was varied from 19,600 to 31,300 revolutions per minute corresponding to a peripheral disc speed of from 106 to 169 m/second. The inlet temperature was constant at 195°C, and the outlet temperature was kept constant at 90°C by gradually increasing the feed rate.

By increasing the degree of atomization the concentrated milk is given a large surface area, so that the water is more readily evaporated. Therefore, a lower moisture content is achieved at the same outlet temperature.

The solubility shows a marked improvement with increased speed of the atomizer even where the powder is over-dried. But low moisture content is not caused by high temperature, as was the case in test two, and therefore, it has no adverse effect on the solubility. The improvement in solubility is due to shortened milk drying time when it is atomized into smaller particles.

One would expect free fat levels to increase due to the increased powder surface area. However, by increasing the speed of the atomizer, a certain homogenizing effect is obtained in the disc, which may explain the modest drop in free fat.

The density is only slightly affected, and decreases slightly due to lower moisture content.

The last experiment (table 6) shows the influence of heating the concentrate. The air inlet temperature was constant at 195°C and the outlet temperature was kept constant at 90°C by gradually increasing the feed rate. The atomizer speed was constant at 25,000 r.p.m.

The effect of heating the concentrate is lower viscosity, and therefore, atomization into smaller droplets. This was achieved in the previous experiment by increasing the atomizer speed, but the result is of course the same, and therefore, there is good agreement between the results in tables 5 and 6 with regard to the changes in moisture content, solubility, and bulk density.

TABLE 6

Influence of Concentrate Heating at Constant Inlet Temperature 195°C, Constant Outlet Temperature 90°C, and Constant Atomizer Speed 25,000 r.p.m.

Non-homogenized Whole Milk Concentrate, 49.5% Solids

concentrate temperature °C	moisture content %	ADMI solubility index	free fat content % of total fat	bulk density g/cm^3
50	3.18	1.20	13.8	0.58
60	2.87	0.60	14.5	0.57
70	3.04	0.30	15.9	0.55
80	3.02	0.20	21.1	0.54

The free fat starts at a higher level, because the concentrate was not homogenized. The increase in free fat by increasing concentrate temperature is due to the increase in powder surface, the effect of which has not been off-set by increased homogenization in the atomizer disc, as in the previous test, because the disc speed was kept constant.

Summarizing the results of the last two experiments the following effects are noted:- When milk is atomized into smaller droplets, the outlet temperature can be reduced and the solubility is improved.

Smaller particles are obtained by increasing the speed of the atomizer or - in case of nozzle atomization - by increasing the pressure.

Smaller particles can also be obtained by reducing the viscosity of the feed, - either by reducing the total solids or by heating the concentrate. Concentrate heating has the highest effect at high concentration. The tendency to increase in free fat content in whole milk powder due to the larger powder surface can be counteracted by increased atomizer speed (or increased nozzle pressure) or by homogenization of the concentrate. Homogenization does, however, increase the viscosity, and therefore, these two contradictory parameters have to be optimized when whole milk is dried.

The experiments clearly illustrate that a slight change in process conditions significantly affects the characteristics of the powder.

ADDED-VALUE PRODUCTS

A development of recent years has been that drying of special products has gained increasing importance to the industry, to get a higher return compared with drying of skim- and whole milk. As costs of milk and labour increase, it is essential to produce products with higher added-value.

Some of these products have been known for a number of years such as instant powder, fat-filled powder, heat classified powder, baby food etc. Products introduced in recent years include whey protein powder and permeate powder. Special technology is required to produce these products. In this section, some important points regarding manufacture of baby food powder are discussed.

Baby Food Powder

Since the middle of the 20th century, emphasis has been placed on the advice from paediatricians that baby food powder should approximate to the composition of breast milk as closely as is practicable. Medical authorities advise against feeding infants younger than four months old with unmodified cows' milk. This has led to the development of the so-called humanized baby food powder suitable for babies from day of birth. From four months of age the infants can be fed on cows' milk that has been only slightly modified.

The total solids of breast milk and cows' milk is about the same, but there is considerable difference in the composition. It is, therefore, necessary to modify cows' milk to obtain a product similar in analysis to breast milk.

However, breast milk is not well defined. It varies widely in composition, particularly during the first 14 days after parturition. In practice, humanized baby foods are produced to match the average analysis of breast milk produced some 14 days after parturition, - the so-called matured milk. This analysis should not be interpreted too rigidly, because, due to natural variations, many mothers produce milk "out of specification".

Table 7 shows the composition of average, matured breast milk and cows' milk. It will be seen that when infants are fed with normal full cream cows' milk instead of breast milk, they will receive a higher protein intake with different composition, different fatty acid composition, less carbohydrate, and a higher mineral intake.

The high content of protein and minerals in cows' milk puts a high load on the infants'kidneys. It is essential that humanized baby foods do not have too high protein content and have low mineral level, as the capacity of the kidney of a newly born child is very limited. This becomes particularly critical when the loss of water is increased by fever or diarrhoea.

TABLE 7

Composition of Breast Milk and Cows' Milk

		Average Matured Breast Milk per 100 ml	Cows' Milk per 100 ml
Total Protein	g	1.5	3.3
Protein Composition:			
Whey Proteins	%	70	20
Casein	%	30	80
Fat	g	3.3	3.5
Fat Composition:			
Unsaturated Fats	%	52	43
Saturated Fats	%	48	57
Carbohydrate	g	7.0	4.7
Minerals	g	0.21	0.72
Energy	kcal	72	66
Protein Calories	%	6	21

The simplest modification of cows' milk is addition of carbohydrate, which will dilute the proteins and minerals.

The protein composition is changed by means of whey solids, whereby the level of lactose is also increased. Whey which has been ultrafiltrated should not be used, because the lactose in the whey will be removed from the protein fraction by ultrafiltration.

The whey must originate from cheese manufacture where additives such as nitrate or sodium chloride have not been used. The whey proteins must be undenatured, i.e. the heat treatment of the original milk and the whey must be gentle, and the whey must be at least 90% demineralized by electrodialysis or ion-exchange.

With regard to the fat, it is possible to change the fatty acid composition of cows' milk by addition of vegetable fat, whereby the level of unsaturated fats is increased.

Addition of lactose is normally used to adjust the carbohydrate level.

As far as the minerals are concerned, it should be mentioned that calcium, magnesium, and phosphorus are desirable in certain quantities, whereas potassium, sodium, and chlorine should be avoided. The low mineral level can be achieved by using demineralized milk as well as demineralized whey.

Breast milk is low in iron and iron deficiency can be avoided by addition of iron to baby foods.

With regard to vitamins, the only addition to those naturally present in cows' milk are Vitamin A, Vitamin D, and Vitamin C.

Possible contaminants in cows' milk such as trace metals, traces of pesticides, herbicides, and antibiotics must be kept under strict control, and it is, therefore, not unusual to specially select the milk to be used for manufacture of infants' foods to ensure absence of contaminants.

In summary the position with regard to infant foods is that modern technology has made it possible to modify cows' milk to produce a baby food which has properties similar to breast milk, and therefore is safe to use from the day of birth. The only shortcoming of the manufactured baby food is that it does not possess the immunological properties of breast milk. It is well known that breast milk has anti-infective properties, and no method exists today to give manufactured baby foods these properties. Therefore, although the manufactured baby foods give the infant all necessary nutrients, breast feeding should always be used where possible.

Great care must be taken in the manufacture of infant foods, to safeguard the bacteriological quality of the finished product. It is not unusual to require a plate count as low as 50 organisms per g of powder with pathogens absent.

Dry mixing of ingredients into the powder after spray drying should be avoided, as it can lead to bacteriological contamination. Further, by dry mixing it is difficult to obtain a uniform distribution, and the added ingredients - e.g. lactose - may segregate out in the cans when they are subjected to vibrations.

Carbohydrate should be added to the milk before evaporation or to the concentrate in the form of a 50% solution, which must be pasteurized before it is mixed with the product.

The fat soluble vitamins (Vitamin A and Vitamin D) should be added to the milk before evaporation and the same aplies to the vegetable fat and possible other ingredients such as dextrin-maltose. The mix should be homogenized, pasteurized at 110°C with one minute holding and evaporated to 47-48% total solids. The concentrate should have a temperature of 60-70°C and should be spray dried using an air inlet temperature of 180°C.

The heat sensitive vitamins (Vitamin C) should be added to the concentrate. It is necessary to overdose with Vitamin C as a portion is destroyed during drying. However, it is more costly to dry mix the vitamin into the dried product, and it will be difficult to obtain a uniform distribution. Further, by adding Vitamin C before drying, the manufacturer can guarantee the end-user that the child receives the correct amount of vitamins because the portion with the lowest heat resistance, which in case of dry mixing may be lost when the powder is reconstituted in hot water, has already been lost in the dryer and substituted by overdosing the concentrate.

Improvement of the reconstitution properties of baby food powder by means of an instantizing process is normally not necessary, because the powder is reconstituted in hot water and the high carbohydrate content gives the powder instant properties.

REFERENCES

1. J. De Vilder, R. Martens and M. Naudts, Milchwissenschaft 31 (1976) 396 - 400.

2. J. De Vilder, R. Martens and M. Naudts, Milchwissenschaft 34 (1979) 78 - 84.

FIRE HAZARDS IN SPRAY-DRYING OF MILK PRODUCTS

E.C. SYNNOTT AND T.C. DUANE*
Dairy and Food Engineering Department, University College, Cork.

ABSTRACT

The occurrence of major fires and explosions during the spray drying of milk product powders in Ireland and similar incidents reported from other countries are outlined. The causes of these problems are looked at. These include self-heating of powder deposits, external sources for heating deposits, mechanical malfunction of equipment and operational hazards such as problems at dryer start-up, and caustic soda contamination of feed. Various protection methods against fires and explosions in industrial dryers are considered.

Theories of self-heating, including Semenov and Frank-Kamenetskii models, the extension of the latter's work to relate sample-size and minimum ignition temperature (MIT) and its application in laboratory testing of powders are assessed. The effect of sample shape on the self-heating characteristics of powder is discussed, and possible errors are pointed out in extrapolating these size-temperature relationships to give estimates of critical layer thickness for deposits in spray dryers in their lower operating temperature regions. Results from oven, hot-plate and flow-through oven tests on the MIT of a variety of milk-product powders are presented. These include the effect on MIT of fat-percentage in the powder and of its total unsaturation. Attempts at using Differential Thermal Analysis techniques to assess the contribution of the type and chemical quality of the fat incorporated in the powder to the lowering of the MIT are outlined.

* Present address:- College of Catering,
Cathal Brugha Street,
Dublin 1.

INTRODUCTION

Spray-drying of milk products is an integral part of the world's dairy industry. The reasons for the existence of the process, details of its technique and specifications for its products are well known and, indeed, the present position in these areas is reviewed in other papers in this symposium.

One of the major risks of spray drying and subsequent handling of products is the possibility of fires and explosions. While these safety problems may not be widely publicised, nevertheless, they can be both hazardous to personnel and costly to the processing company and its insurers, in terms of plant damage and shut-down while repairs or replacement takes place. The Institution of Chemical Engineers (ref. 1) quotes one explosion and over thirty fires per annum in the U.K. in recent years for which a local authority fire brigade had to be called. In Ireland, in a recent 10 year period, approximately 10 major incidents occurred in spray dryers where damage exceeded IR£100,000. The maximum bill was of the order of IR£2m.

The twin factors of personnel safety and the cost of down-time and repair indicate that the risk of these outbreaks occurring must be minimised by both the designers and operators of spray-drying plant.

It is proposed to divide this paper into two parts:-

(a) a commentary on incidents that have occurred and the techniques used in this country to counteract them, and
(b) laboratory testing of milk product powders to assess their potential to cause fires.

Cause of fires/explosions

The causes of various incidents have been investigated in many countries, and work by the authors with the Irish Dairy Industry has indicated the following sources of trouble: -

1. Friction Heating - between rotary valve and cyclone housing
2. External Heating - welding of ductwork causing ignition of deposits inside
3. Equipment Malfunction

   (i) Air disperser coming adrift
   (ii) Atomiser spindle broken
   (iii) Spark from heat source into dryer
   (iv) Powder into air disperser due to design fault
   (v) Overheating in inlet area due to cooling fan failure
   (vi) Interlock failure and powder into inlet from fines return pipe.

4. Self Heating of Deposits

   (i) On air disperser vane, ignited and fell off into main chamber
   (ii) On air disperser, ignited and sucked into fluidised beds
   (iii) On fines return pipe
   (iv) Babyfood deposits in dryer
   (v) Contamination of feed with caustic soda.

5. "Snowballs" - round nodules of powder which grow in size, particularly in fluidised beds, self-heat and disintegrate when they hit the wall.

6. It has also been noted, in common with other investigators, that many fires occur within one hour of restart with product after a clean-up. A number of suggestions have been made to account for this:-

1. Product deposits remaining from previous run
2. Non-steady-state conditions at start-up

3. "Sticky" powders causing thicker deposits
4. Wet dryer at start-up causing thicker deposits
5. Soaking of deposits with caustic soda.

Further reference will be made to the significance of some of these factors later.

An analysis of the fire incidents which have taken place in Ireland shows that roughly half occur with skim milk and half with fat-filled powders, even though it is thought that the latter are potentially more dangerous products to dry. This may be explained by the fact that approximately 10 times as much skim powder as fat-filled was manufactured in the period under consideration.

Protection against fires and explosions:

Various techniques are available minimise the effects of fires and explosions. Among methods used in Ireland are:-

1. Use of water sprays to cool the burning material. These are fed through the feed atomiser or spray devices in the roof of the dryer. These may be triggered by heat detectors with a set temperature-differential value. In addition, the feed supply and heat source are disconnected.
2. Pressure-sensitive devices, with a short response time, are used to sense the initial pressure rise associated with a developing explosion to enable a suitable suppressant to be injected very very rapidly.
3. Rapid-acting isolating valves are used between sections of the drying plant to prevent the carryover of the problem from one section to another.
4. Since the spray-dryers used in the Dairy Industry cannot be economically manufactured to contain an explosion, venting is required for any pressure rise associated with fire or explosion. This venting must be done very rapidly through suitably sized, designed and located pressure relief doors to prevent structural damage to the plant and to allow the products of the explosion to be safely discharged to the outer air. Vents are usually installed in the roof of dryers but

may take the form of hinged doors in the walls. These should be flush-fitting so that dust cannot accumulate on ledges. All vents should lead to the outside of the building and there is a limit on the length of ducting which may be used if the pressure rise in the chamber is to be limited. When an explosion occurs, burning dust is vigorously expelled through the vent and this fact must be considered in locating the vent point.

Calculation of vent area required

Much has been written concerning the explosion relief area needed in a milk product spray dryer. Various international standards and criteria are quoted which give appreciably different values for the area required. Beever (ref. 2) gives figures for vent area required for milk product spray dryers based on a number of these criteria:-

(i) Vent Ratio $= \dfrac{\text{Area of vent}}{\text{Volume of vessel}} = \dfrac{A}{V}$

The appropriate ratios are given in Table 1

TABLE 1
Relationship between Spray-Dryer Volume and Vent Ratio

Volume m^3	Vent Ratio m^{-1}
<30	1 : 6.1
30-300	1 : 6 to 1 : 1.25
300-600	half top area
>700	full top area

It is assumed that the plant can withstand an overpressure of 0.14 bar. These values of vent ratio, when applied to the large spray driers in use in Ireland, give large vent areas which are difficult and costly to accommodate. For vessels which are designed to withstand a maximum pressure rise of 0.2 bar, smaller vent areas may be used.

(ii) Cube Law: Vent area $\alpha V^{2/3}$

This has a certain logical basis since $V^{2/3}$ has the dimensions of area. The value of the proportionality constant depends on the explosibility of the dust, and milk powders are classed as weakly explosible.

(iii) Research carried out by Pineau (ref. 3) in a 100^3m chamber using whole milk powder and measuring the pressure rise as a function of vent area, indicates that smaller vent areas are appropriate, since, in practice, it is unlikely that the entire volume of the dryer contains an explosible concentration of dust.

(iv) "Dirty" Volume

Only that portion of the total volume of the chamber containing an explosible concentration is considered. In the case of a spray dryer, this is the cone volume and leads to a very much decreased vent area.

Exothermic Reaction

When a body of material is capable of self-heating through exothermic reaction between its constituents and oxygen, then the

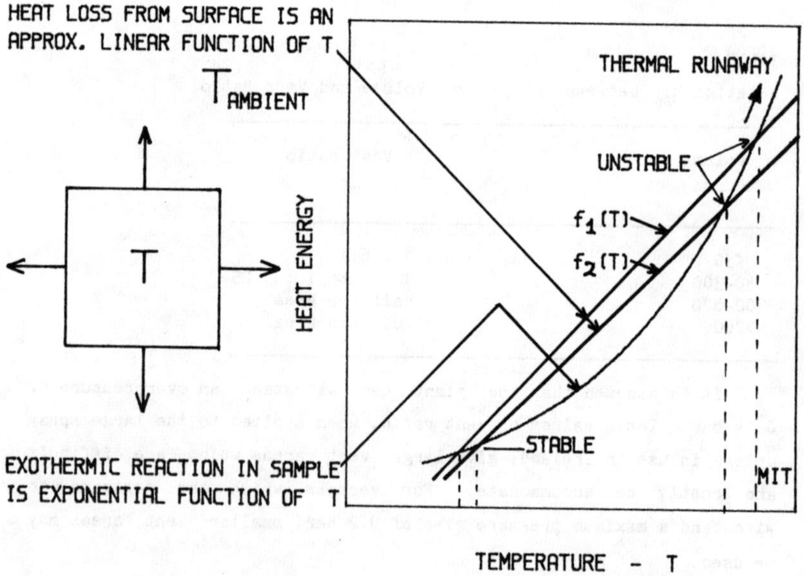

Fig. 1 Typical curves for the heat generated in a body undergoing an exothermic reaction and the heat lost from its surface.

potential for thermal runaway exists. Fig. 1 shows a body of material at temperature, T, where the heat generating reaction is an exponential function of T and the heat lost to the surroundings is an approximately linear function of T. Stable and unstable intersection points exist. If the unstable value is exceeded, thermal runaway occurs. This value is taken as the Minimum Ignition Temperature. Changing ambient conditions (e.g., air velocity over reacting body) may influence the minimum ignition temperature as a result of a higher or lower intersection point.

Laboratory Assessment of Fire Risk with Milk-Product Powders

In laboratory studies of the parameters which influence ignition, use is made of a number of oven and hot-plate designs. The purpose is to find the temperature which will cause a body of powder to self-heat and thus obtain an indication of the safe temperatures to which powder may be subjected in practice.

The shapes of sample used for test purposes namely cubes, layers, spheres and cylinders are amenable to theoretical analysis.

The fundamental theories of self-ignition have been largely based on the models of Semenov and Frank-Kamenetskii. These differ in that Semenov postulates that the thermal conductivity of the material is very high while Frank-Kamenetskii assumes that it is extremely low. These assumptions are summarised in Fig. 2, showing temperature distribution throughout the sample.

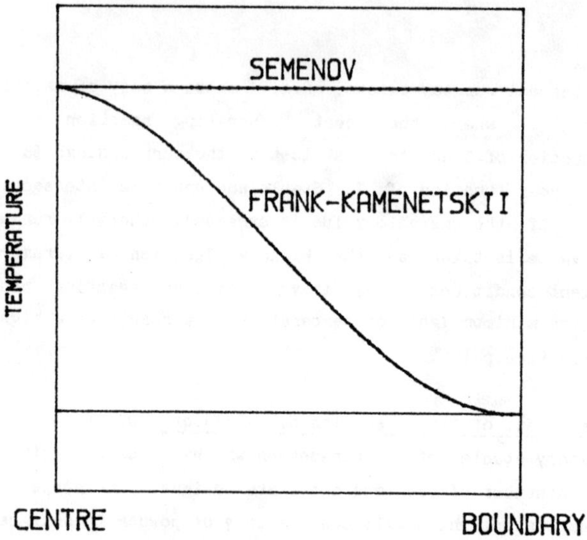

Fig. 2 Temperature distribution in a body undergoing an exothermic reaction according to the theories of Semenov and Frank-Kamenetskii.

Research carried out by many workers including work done in the authors' Department, and summarised by MacCarthy (ref. 4), has shown that the thermal conductivity of milk powders is of the order of 0.036 to 0.11 W/mK and therefore much nearer to the F.K. model than the Semenov. In this project, use has been made of the F.K. model (ref. 5). His basic theory and modifications introduced over the years have produced an equation relating the critical size and shape of a mass of powder to its Minimum Ignition Temperature (ref. 6).

$$\ln \frac{\delta_c T^2}{r^2} = M - \frac{N}{T}$$

where T is the Minimum Ignition Temperature,

r is half the critical dimension of the sample,

δ_c is the F-K parameter, depending very largely on sample geometry,

N + M are constnats from the linear relationship.

N may be shown theoretically to be proportional to the activation energy of the combustion reaction of the sample.

TABLE 2

Typical values of δ_c for the shapes tested

Shape	δ_c
Cube	2.64
Layer	0.92
Sphere	3.46
Cylinder	2.89

Minimum Ignition Temperature (MIT) is defined as the average of the lowest ambient temperature at which a particular size and shape of sample ignites and the highest at which it fails to ignite. MIT may not be sharply defined, as research in the authors' laboratory has on occasions given no overlap between ignition and non-ignition tests, while with some products it may be as high as 10°C.

Test Regimes

I. Static Oven: typical design is shown in Fig. 3. This type of oven gives the MIT for all-round heating of sample.

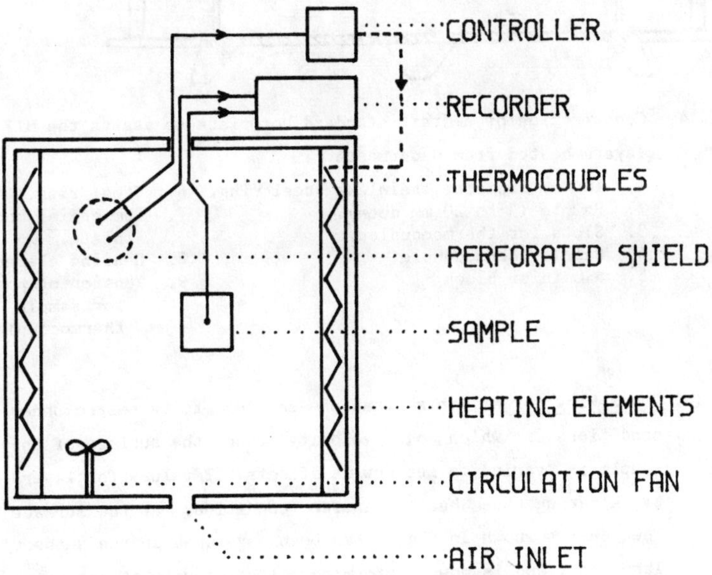

Fig. 3 Layout of a static oven to assess the MIT of a sample under all-round heating.

II. Hot-plate: this design (Fig. 4) has been recommended by Committee 31H of the International Electrotechnical Commission (ref. 7) and simulates the situation where a deposit of powder is laid down on a hot surface (such as a steam pipe or overheated bearing) and is cooled on its other sides by air at ambient temperature. This test gives a much higher value of MIT than an oven test on a similar layer.

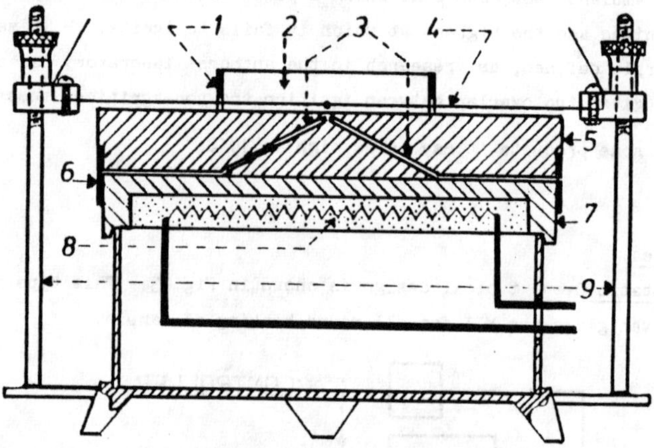

Fig. 4 Construction of a draft-standard hot-plate to assess the MIT of layers heated from underneath.

1. 100 mm diameter stainless steel ring.
2. Sample (5 to 20 mm deep).
3. Slots for thermocouples.
4. Sample thermocouple.
5. Aluminium block.
6. Steel ring.
7. Commercial hot plate.
8. Heating element.
9. Tensioning posts for sample thermocouple.

III Cross-Flow oven: this enables samples to be tested under conditions in which air velocity over the surface of the sample is varied to see how it affects MIT values for layers by altering the heat transfer conditions at the surface. The design shown in Fig. 5 has been developed in the authors' laboratory and is now performing satisfactorily at operating temperatures up to 400°.

IV. Flow-through oven: in this design air at the test temperature is fed up through the sample as in a fluidised bed (Ref. 1). This air could have two conflicting influences on MIT. Firstly, it acts as a coolant on the reaction, thus raising the required MIT. Secondly, it provides an ample supply of oxygen for the reaction thus possibly lowering the MIT. To date, this laboratory has not succeeded in developing a design which is able to withstand the high temperatures reached.

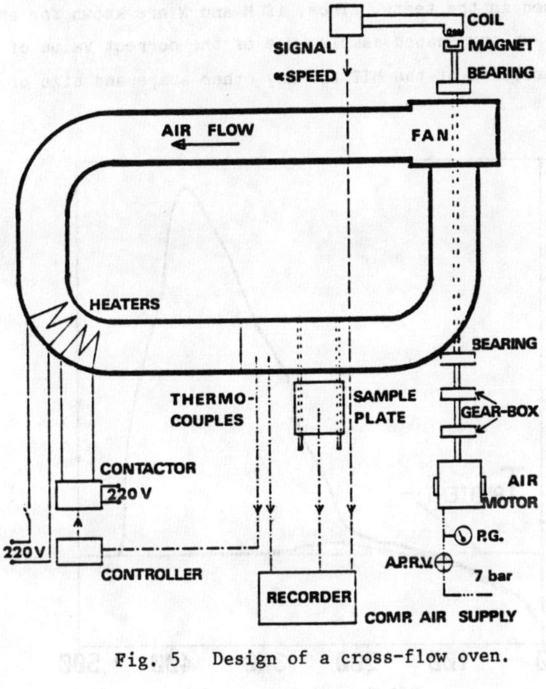

Fig. 5 Design of a cross-flow oven.

Fig 6 shows typical temperature-time response curves for two tests in which the oven temperature differed by 1°C between the two tests. The dotted curve shows the case in which the sample underwent sufficient exothermic reaction to raise its centre temperature above the ambient but not sufficiently to sustain the reaction to thermal runaway conditions. The additional 1°C in ambient temperature was enough to cause the sample to ignite and reach a centre temperature of the order of 750°C.

RESULTS

It has been shown in both static-oven and hot plate tests that the F-K equation fits excellently the data obtained for temperature versus critical size for all milk powder products and sample shapes tested (ref. 8).

It has been maintained by a number of authors that if the appropriate value of δ_c is used, the points on the F-K plot for all shapes and sizes of sample should lie on the same straight line. The reasoning behind this is that the values of M and N depend only on the powder used in the tests. Thus, if M and N are known for any powder from tests on any shaped sample, use of the correct value of δ_c will allow calculation of the MIT for any other shape and size of deposit.

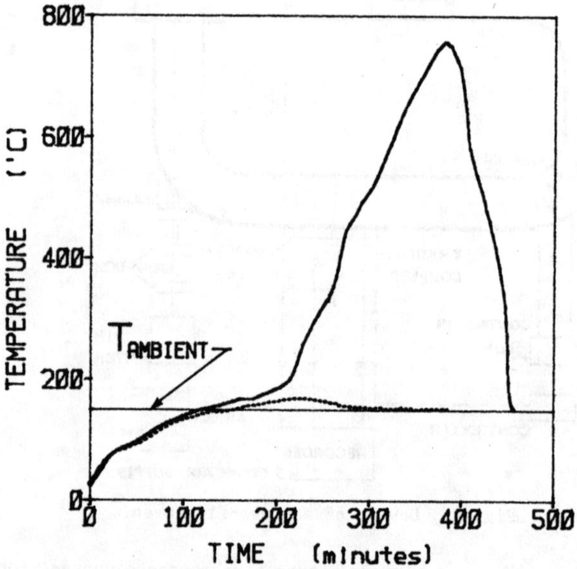

Fig. 6 Typical response curves for 50 mm cubes of 25% cocoanut/1% tallow-filled powder in an oven where one sample ignited and the other did not. The ambient temperature differed by 1°C between the tests.

It was decided to evaluate this hypothesis by testing a 30% tallow-filled powder over a range of sizes (2r):

Cubes	25	38	51		mm
Layers (100 mm in diameter)	10	15	20		mm
Spheres	36	50	77	102	mm
Cylinders (with d=h)	51	76	100		mm

An average of 10 tests was used to assess each MIT and the F-K equation gave excellent straight line fits as evidenced by the coefficients of correlation obtained for each shape (0.988 to 0.999) (Fig. 7).

It may be noted from Fig. 7 (ref. 9) that while the results for spheres and layers coincide, those for cubes and cylinders diverge. While this divergence may not be too significant within the

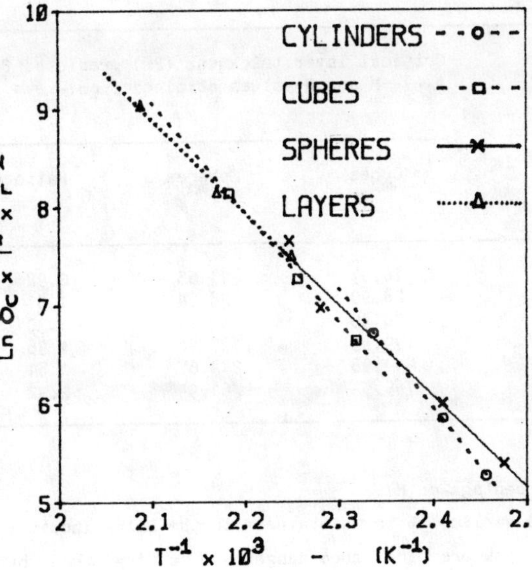

Fig. 7 Plot of ln $\frac{\delta_c T^2}{r^2}$ against $\frac{1000}{T}$ for various shapes and sizes of powder sample. Best-fit straight lines are included for each sample shape.

range considered, it is of major importance when results are extrapolated (as Beever does (ref. 10)) to give estimates of critical layer thickness in lower temperature regions of the spray dryer and to predict safe storage conditions in silos. Results from cube tests have been used for this purpose to date, and a comparison is shown (Table 3) for the results obtained by using values obtained in this project from cube and sphere tests. A very significant difference may be noted between the predictions at low temperatures and this should indicate caution in placing too much reliance on results from any one shape of test sample.

TABLE 3

Critical Layer Thicknesses (2r) and their ratio for selected temperatures predicted from M and N values obtained from cube and sphere tests.

Ambient Temperatures ^{o}C	Critical layer thickness (2r) predicted from M and N values obtained from:-		
	Cubes mm	Spheres mm	Ratio
200	10.57	11.55	0.92
160	28.90	28.54	1.01
130	70.45	54.39	1.30
100	199.43	127.48	1.56
80	441.95	244.67	1.81
50	1765.54	761.94	2.32

Effect of fat percentage on MIT

As mentioned earlier it is maintained in the Dairy Industry that fat-filled milk powders are more dangerous to dry than others, and evidence is available to support this assertion. To elucidate this observation, tests were carried out on a large number of powders, containing a wide range of dairy and other fats, to see if MIT decreased with increasing fat content. Results presented in 1981 by Duane and Synnott (ref. 11) and plotted in Fig. 8, show a wide scatter

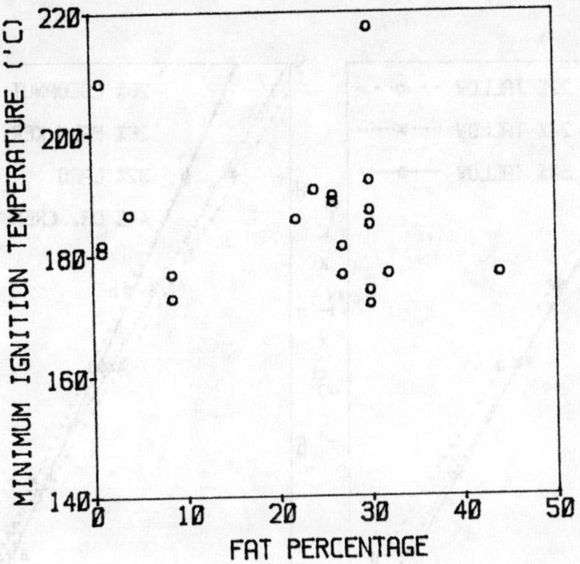

Fig. 8 Plot of MIT against fat percentage for powders with a variety of dairy and non-dairy fats incorporated.

of results and no evidence to support the contention that an increase in fat % leads to a decrease in MIT. It should be noted, however, that the powders were manufactured in a variety of commercial and pilot plants under a range of conditions.

It was decided to examine a limited number of commercially manufactured powders containing a variety of fats. Fig. 9 (a) (ref. 12) shows F-K plots for these powders containing different percentages of tallow, which were tested as cubes of various sizes. Fig. 9 (b) shows results for four powders containing 26% cocoanut, 26% palm kernal, 32% lard and 44% butterfat. It may be noted that the values of the slope of the lines (N) vary considerably, N being proportional to the activation energy of the combustion reaction. If MIT is plotted against fat percentage, there is a tendency towards a reduction in MIT with increasing fat percentage (20-50% fat: MIT 190 - 174°C) while the N vs % fat plot shows an increase with increasing fat percentage. Values of linear correlation coefficients obtained for 7 points are 0.76 and 0.86 respectively.

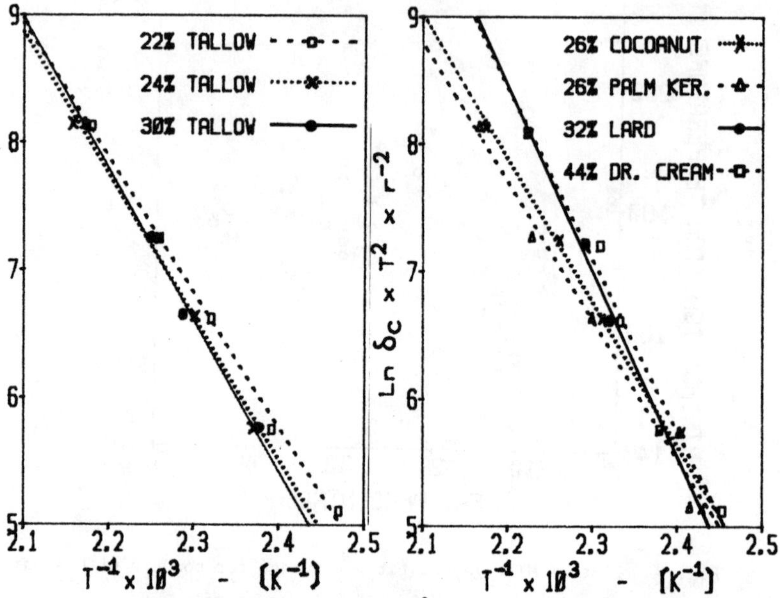

Fig. 9 (a) and (b). Plot of $\ln \frac{\delta_c T^2}{r^2}$ against $\frac{1000}{T}$ for seven powders containing different fat types and percentages.

It was decided to examine the effect on MIT of unsaturation of fat, based on the fact that combustion is an oxidative reaction, and unsaturation is known to predispose fats to oxidation. The Iodine Number (a measure of unsaturation in lipid material) was multiplied by the fat fraction to obtain a proxy for total unsaturation. MIT and N were plotted against total unsaturation, and results showed the same trends as before - decrease in MIT and increase in N with increasing total unsaturation. The coefficients of correlation, however, decreased slightly to 0.59 and 0.81 respectively.

The possibility of using Differential Thermal Analyis (D.T.A.) techniques to assess the response of milk powders to controlled heating rates was next investigated. In this method small samples (10-15 mg) of the test powder and an inert reference material of similar thermal properties are placed in containers on thermocouples in a furnace (Fig. 10) where temperatures may be

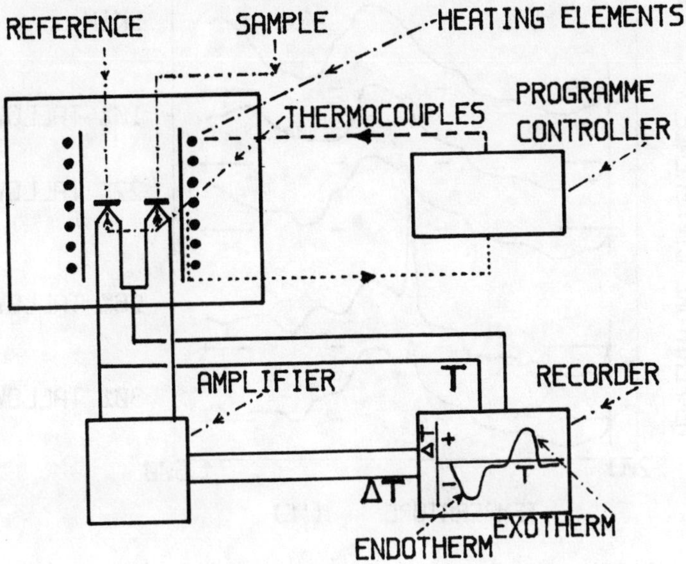

Fig. 10 Schematic diagram of a Differential Thermal Analyser showing a typical ΔT versus T curve.

raised at a fixed rate from ambient to 500°C. Thermocouples are used to measure directly the temperature difference between the test powder and the reference material. Typical responses are shown (Fig.10) for endothermic (heat-uptake) and exothermic (heat generation) reactions. From tests on milk powder constituents, the indications are that the protein fraction is the main reacting material producing exotherms. Typical responses are shown (Fig. 11) for skim powder and a number of tallow-filled powders (ref. 13). Above 10% tallow filling, a characteristic spike appears on the DTA trace and the area and amplitude of this spike varies.

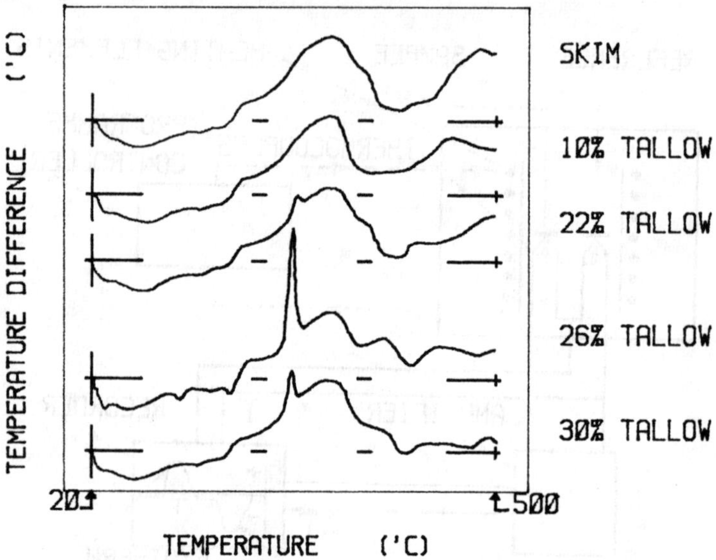

Fig. 11 DTA curves for skim powder and four tallow-filled powders.

A variety of tallow-filled powders was tested for which analytical data was available on total unsaturation, quality of fat, etc. Tests were also conducted on fat extracted from the powder. (Ref. 13). It appears that the characteristics of the added fat may play a part in determining ignition characteristics of the powder. The characteristic spike with tallow-filled powders has not been observed with other added fats.

Cross-flow Oven

A 25% cocoanut/1% tallow-filled powder was used to assess the effect of varying air velocities on the MIT of layers 15 mm deep and 100 mm diameter in the cross-flow oven. Fig 12 (ref. 13) shows the results obtained, and it is clear that air velocity has an effect on MIT, which decreases from 230°C to 215°C as air velocity increases to 1.5 m/s. Thereafter, a very gradual decrease takes place with increasing velocity up to 9 m/s. One implication of these results is that deposits in spray-dryers exposed to a high air velocity may present an increased ignition risk.

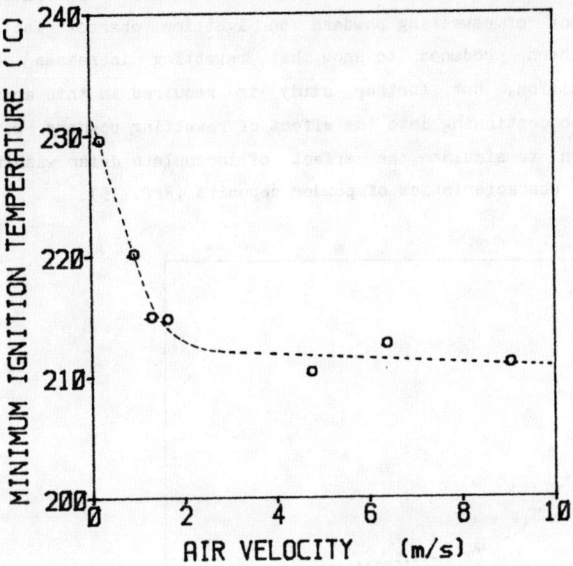

Fig. 12 Effect of air velocity in a cross-flow oven on MIT of 15 mm layers (100 mm diameter) of 25% cocoanut/1% tallow-filled powder.

Role of Caustic Soda in Self-ignition of Milk Powders

Investigation of a minor fire in a processing plant made it clear that glowing deposits in the main drying chamber were associated with contamination of the milk concentrate with caustic soda, which occurred when a malfunction in an isolating valve between process and cleaning circuits resulted in the addition of caustic soda to the feed. Accordingly, tests were carried out on powder which had been manufactured just before the fire incident and on contaminated powder. It was found for 25 mm cube samples that the MIT of the contaminated sample was 160.5°C, compared with 176.5°C for control sample (Ref. 14).

Skim powders contaminated with different percentages of caustic soda were manufactured in a laboratory spray dryer. Fig. 13 shows that the MIT decreases rapidly with the addition of small percentages of caustic, showing the danger of even slight caustic contamination.

Investigations have been carried out in the authors' laboratory into the effect of rewetting powders on ignition characteristics. Evidence has been produced to show that rewetting increases the tendency to ignition, but further study is required in this area. Studies are also continuing into the effect of rewetting powders with caustic solution, to simulate the effect of incomplete drier washing on the ignition characteristics of powder deposits (Ref. 15).

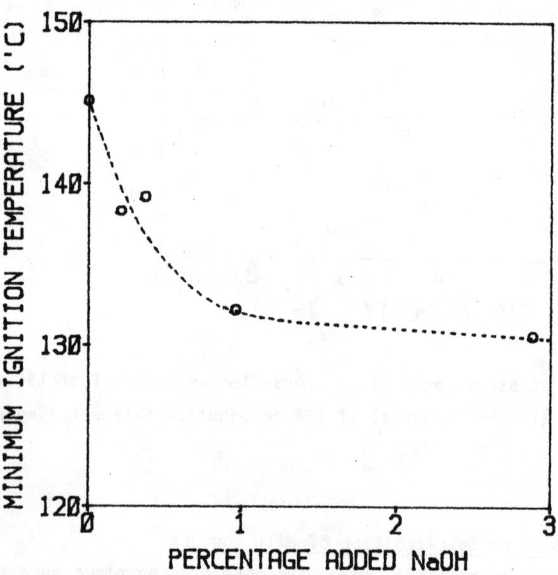

Fig. 13 Effect of percentage caustic soda incorporated in skim powder on the MIT of 50 mm cubes in a static oven.

CONCLUSIONS

The fire/explosion incidents which have taken place to date are not confined to any single type of dryer or powder, although a disproportionate number appear to be associated with manufacture of fat-filled product. Increasing fat level and level of total unsaturation of the fat tend to result in decreased MIT values. An increase in air velocity over a deposit produced a similar effect. The protein fraction seems to be the component responsible for the major exothermic reaction in self heating but DTA evidence would suggest that the ignition properties are modified by the type and condition of added fat, particularly if that fat is tallow. Finally, the role of caustic soda contamination in reducing MIT is quite significant and the effect of soaking deposits with caustic may be a viable explanation for a number of hitherto unexplained fires.

ACKNOWLEDGEMENTS

The authors wish to thank the ten powder manufacturers and the National Board for Science and Technology who contributed to the funding of research in this subject area. They also express their appreciation of the technical assistance received from Messr. J. O'Mahony and J. Barrett.

REFERENCES

1. User Guide to Fire and Explosion Hazards in the Drying of Particulate Matter, Institution of Chemical Engineers, Rugby, 1977.
2. P. Beever, Fire and Explosion Hazards in the Spray Drying of Milk, Paper VI, Developments in Milk Powder Technology, 6 December, 1984, Portlaoise, Agricultural Institute, Dublin, 1984.
3. J.P. Pineau, Protection against Fire and Explosion in Milk Powder Plants, Proc. 1st International Symp., Explosion Protection in Practice, Antwerp, April, 1984.
4. D. MacCarthy, in Engineering and Food, Vol 1, Dublin, September 26-28, 1983, Elsevier Applied Science, London, 1984, pp 527-538.
5. D.A. Frank-Kamenetskii, Diffusion and Heat Transfer in Chemical Kinetics (translated by J.P. Appleton), Plenun Press, London, 1969, pp 374-421.
6. P. C. Bowes, A general approach to the prediction and control of potential Runaway Reactions, Proc. Symp., Runaway Reactions, Chester, 24-26 March, 1981, Institution of Chemical Engineers, Rugby, 1981, pp 1/A: 1-35.
7. Draft Standard prepared by Committee 31H, Working Group 2, International Electrotechnical Commission, November, 1976.
8. T.C. Duane & E.C. Synnott, Comparison of Minimum Ignition Temperatures obtained from oven and hot-plate tests on a fat-filled Milk Powder. Brief Communications, 1, Book 2, XXI International Dairy Congress, Moscow, 12-16 July, 1982.
9. J.G. O'Mahony and E.C. Synnott, Influence of Sample Shape and Size of Self-Ignition of a fat-filled Milk Powder, submitted for publication, J. of Food Engineering
10. P. Beever, Spontaneous Ignition of Milk Powders in a Spray Drying Plant, J. Soc. Dairy Tech. 37 (2), 1984, pp 68-71.
11. T.C. Duane and E.C. Synnott. Effect of some physical properties of milk powders on Minimum Ignition Temperature, Proc. Symp., Runaway Reactions, Chester, 24-26 March, 1981, Institution of Chemical Engineers, Rugby, 1981. pp 2/J: 1-8.

12. T.C. Duane & E.C. Synnott. Ignition Characteristics of Spray-Dried Milk-product Powders in Oven Tests, prepared for publication.
13. D.Twomey & E.C. Synnott. Research project in progress.
14. J.G. O'Mahony and E.C. Synnott. Research project in progress.
15. M. Everard and E.C. Synnott. Research project in progress.

ADVANCES IN THE DEHYDRATION OF FRUITS AND VEGETABLES

S.D. HOLDSWORTH
Campden Food Preservation Research Association, Glos.

ABSTRACT

This paper reviews the developments in the dehydration of fruit and vegetables with particular reference to the processes used in the production of rapidly reconstitutable solid products as well as powdered products.

INTRODUCTION

Dehydrated products have the advantage over other forms of preserved foods that they are easily packaged and stored at ambient temperature conditions. However, conventionally air-dried products have tended to be difficult to rehydrate due to structural changes in the product and of indifferent quality due to excessive thermal damage. During the last two decades, much research has been undertaken to improve the rehydration times and to improve the quality of dried products especially fruits and vegetables. Two different approaches have been taken: one, to improve heat transfer, lower the process temperatures and expand the porous structure of the product and two, to use non-thermal methods or additives.

Table 1 shows how dried fruits and vegetables are utilised in the production of other products. Conventional grape products are widely used for cakes, in sponge pudding formulations and biscuits as well as constituents in dry "health food" mixtures; other fruit products are used in yoghurts, preparation of pie-fillings, components in formulated food products. Fruit juice powders, usually deliquesent and requiring very good packaging are widely used for making fruit juice drinks.

The market for dried vegetable products has increased greatly over the past two decades. "Instant" potato products, granules and powders have become particularly prominent, and more recently rapidly reconstitutable dried vegetable mixtures often containing rice or pasta products have become important, as well as dried soups both instant and conventional.

This review covers some of the developments which have taken place during the last two decades with regard to the processes which have been developed for achieving the new products. It has been necessary to be selective because of the vast amount of literature and patents which have been produced during the period.

PROCESS DEVELOPMENT

General Aspects

The literature and patents up to the middle '70s have been reviewed in a number of publications (refs. 1-6). During this period freeze drying and the less expensive explosion puffing were developed in order to produce products which could be rapidly reconstituted, ie. in less than 5 minutes, and which could be used in formulated meals as well as fast food production. High efficiency band driers were also developed which were flexible and conserved energy. Fluidised bed drying was also developed and used commercially for producing a range of high quality dehydrated vegetables.

During the last decade the objectives have expanded to include utilization of solar energy, microwave energy and the use of low temperature processes, eg. osmotic dehydration as well as the development of low energy processes for developing countries and for utilizing tropical crops. A review of some of these developments has been published by Setty and co-workers (ref. 7).

Table 2 summarises some of the processes which have been used for producing dehydrated fruit and vegetable products.

Osmotic Dehydration (refs. 7-27)

This process consists essentially of placing the food product in a solution of sugar or salt and allowing the water to diffuse from the product. The quantity of moisture removed depends upon the concentration and temperature of the osmotic solution, the time of immersion and the nature including area of the exposed surface. A typical result for a 60% sugar syrup at 50°C is the dry matter increases from 15% to 46% after 6 hours, and under vacuum conditions even higher concentration can be achieved. The final stage in moisture removal is usually air drying, but other techniques such as microwave drying are feasible.

Most workers have used sucrose up to 70 deg. Brix, although glucose and isomerised glucose-fructose have been used (11). Cuticular wax and wetting problems have been found to be overcome by pre-treatments (8) eg. blanching, or treatment with ethyl oleate (9) or sodium hydroxide (12). Experimental results for a wide range of products have been presented, cherries (8, 9, 11); plums (8, 9, 12); grapes (8, 9); apples (12, 17, 19, 21, 26, 27); pears (11, 12, 13, 26); peaches (12); apricots (11, 26); pineapple (13); banana (13, 19, 25); strawberry (13); papaya (14, 18); Guava (15, 25); mango (18, 19, 20, 25); pawpaw (18, 25); melon (25); and quince (26).

Methods of increasing the osmotic effect have been studied. These include increasing the solute concentration and the addition of acids such as acetic, citric, lactic and hydrochloric acids (18). With regard to the former, the higher the concentration the greater the amount of water removed. However, the latter depends upon the product. For example, acidification of the sucrose increased the loss of water from papaya but not mango.

Relatively little work has been carried out on vegetables using brine although some success has been achieved with okra, aubergines, tomato and peppers (10) and potatoes (22).

Pneumatic Drying

A method of producing quick-cooking vegetables is to use high temperature short time pneumatic ring drying of the dried product as a preliminary to conventional air drying (28). This results in the development of rigid porosity which enhances the drying rate and subsequently the rehydration process. This type of process is only applicable to high starch products eg. potatoes, peas, carrots, sweet potatoes, since it is essentially the starch cell structure which is expanded and produces a satisfactory porous structure. Low starch vegetables eg. pepper, aubergines, French beans were shown not to respond to this kind of treatment. Typical conditions of drying are 160-180°C for 8 min prior to conventional drying at 60-70°C.

Freeze Drying

Most of the recent work has been directed towards the freeze-drying of fruit juices (refs. 29-43). The main problems are with the freezing operation because of the low temperatures required for solidification (ref. 32), retention of volatiles (ref. 42) and the hydroscopic nature of the product (refs. 29, 30, 36). Interest has been shown in vacuum puff freeze drying (refs. 29, 30, 36), which involves the production of a frozen liquid or puree in a foam by application of vacuum followed by conventional freeze drying. The technical problems of hygroscopicity have been solved to some extent by the addition of sucrose (refs. 31, 36, 43). Economic feasibility studies (refs. 29, 30) show that the process is viable in relation to a high quality fruit nectar powder for retail consumption.

The process which has attracted attention in the vegetable dehydration area has been compression freeze drying, which involves simultaneous compaction and freeze drying under vacuum (ref. 44) This increases the bulk density from 0.05 gm/cc to 0.5 gm/cc. For diced carrot or peas a pressure of about 1.5 kg/cm^2 is applied during the freeze drying cycle to the pre-frozen product. The final product is a disc shape and requires boiling in water for 15 mins to rehydrate followed by a simmer of 15 mins at 85°C.

Atmospheric freeze drying has also received attention (ref. 45). Diced carrots and potatoes were frozen and then dried with an airblast 1.2 m/s at 70°C; the time taken for the drying was 120 min.

Solar Drying

During the last decade there has been considerable interest in the use of solar energy (refs. 46-56). The equipment has ranged from simple box-like enclosures with differing covering materials (ref. 47, 49, 51) to forced air convection driers using optical solar concentrators (refs. 55, 56). All these devices have shown significant improvements in quality and reduction in drying time when compared with direct atmospheric sun drying. Solar dehydrators are well suited to rural conditions and small scale food processing industries. A typical result from the drying of green peas or potato products is that moisture was reduced to 6% in 7 hours using a solar heated enclosure which reached 25°C. Some more novel systems which have been examined are a fluidised bed grape drier using solar heated air (ref. 48), and the use of desiccants (eg. silica gel, activated alumina, anhydrous calcium sulphate and molecular sieves) which could be regenerated by solar heat (ref. 51).

Spray Drying

This well-established technique still has considerable potential in applications to drying fruit and vegetable products. An interesting study has been carried out on tomato products (ref. 57) in which the problems of drum-drying and high-temperature spray drying of tomato paste have been discussed in relation to the SPREDA low temperature spray drying system. This uses air at 50°C with a preselected moisture content. The resulting powder is further dried using a fluidised bed.

Other techniques

Energy use and thermal losses associated with tunnel dehydrators have been studied in terms of minimising air leakage, increasing air recirculation and using heat shields (ref. 58).

A pilot plant double drum drier (ref. 59) has been described for studying the drying of fruit purees. It has been shown to have successful application to high fibre products eg. apple, banana, apricot, papaya and cranberry.However low fibre products such as raspberry, blueberry and strawberry require the addition of low methoxyl pectin. Drum drying of fruit pulp, apples, apricots and blackcurrants has also been studied (ref. 59) and appropriate conditions determined viz. temperature - 110-115°C; film thickness - 1mm; drying time - 3 mins; final moisture content - 5%. Drum drying of banana puree has also been studied (ref. 60) in relation to film formation and use of emulsifiers. It was found that monostearin and lecithin both gave excellent film structure.

SUMMARY

From this survey of processes used for the preservation of fruit and vegetable products there are several useful indicators for future development. The most promising of these are the further development of processes to produce rapidly reconstitutable products,in particular belt trough drying and spouted-bed drying. The next decade will see the development of electrical energy heating in particular the re-application of microwave heating as part of a combination technique.

TABLE 1

Utilization of Products

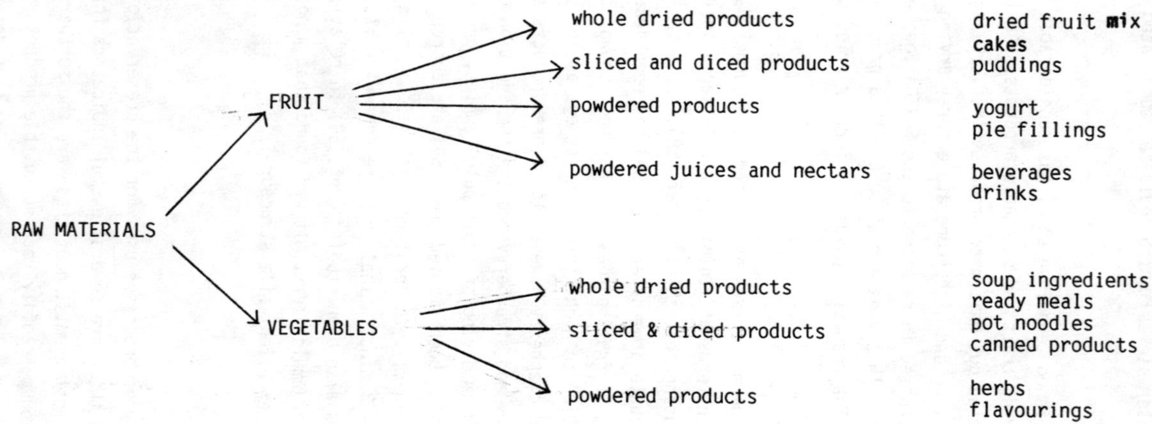

TABLE 2

Comparison of processes

TECHNIQUE	APPLICATION	COMMENTS
1. BATCH TRAY CABINET DRYING	Vegetables: Peas, beans, potatoes, carrots, onions Fruit: Apples	Low maintenance cost Inflexible Poor control Slow production rate Variable product quality
2. TUNNEL TRAY DRYING	See 1 above	As 1 above but more flexible and faster production rate Largely replaced by belt and other types of drier Slow rehydration
3. CONTINUOUS BELT DRYING	Wide range of vegetables and some fruits Apples	Good control High production rates More uniform quality Slow rehydration
4. BELT TROUGH DRYING	Limited to particulate products: diced potato, peas, sliced beans, dried onions & peppers	Good control More uniform drying rate More efficient use of energy Uniform quality Slow rehydration Unsuitable for sticky products
5. FLUIDISED BED DRYING	Limited to products which will fluidise uniformly and create stable conditions	As 4 above but will handle high moisture products High drying rates Uniform quality

TABLE 2 (Continued)

Comparison of processes

	TECHNIQUE	APPLICATION	COMMENTS
6.	CENTRIFUGAL FLUIDISED BED DRYING	As 5	High air velocities Lower temperatures Uniform quality
7.	SPOUTED BED DRYING	Diced products Larger than 5	Good control High drying temperatures High degree of mixing
8.	PNEUMATIC AND RING DRYING	Limited to granules, powders and some low density products eg. parsley Widely used for potato granules	Low residence time High throughput Good control High thermal efficiency
9.	EXPLOSION PUFFING	Diced potato, carrot and some fruits	Intermediate process with conventional air drying Products rehydrate rapidly
10.	FREEZE DRYING	Wide range of products including heat sensitive fruits	High cost process High quality - low heat damage Rapid rehydration
11.	VACUUM DRYING	Wide range of products particularly useful for heat sensitive fruits	High cost process High quality Low temperature Slow production rate Loss of volatiles
12.	MICROWAVE DRYING AND VACUUM	Wide range of products especially high moisture and heat sensitive products	High cost process Good heat transfer Rapid internal heating High quality

TABLE 2 (Continued)

Comparison of processes

	TECHNIQUE	APPLICATION	COMMENTS
13.	OSMOTIC DRYING	Useful for heat sensitive fruit products, soft fruits, peaches, apricots, plums and other drupes	Slow process Low temperature process High quality product
14.	AZEOTROPIC DEHYDRATION	Vegetable dice fruit products	Complex expensive process Difficult to achieve commercially
15.	DESICCANT DRYING	Small particles only	Low temperature Very slow process Rapid reconstitution time
16.	DRUM DRYING	Wide range of liquid pastes for production of potato, tomato and fruit powders	High drying rate Large volume production Good quality
17.	SPRAY DRYING	Used for heat sensitive products	Rapid drying Large volume production Good quality
18.	VACUUM BAND DRYING	Wide range of products especially used with foamed products	Expensive Fast drying Slow production Rapid rehydration Good quality
19.	FOAM MAT DRYING	Any liquid or pureed fruit or vegetable product	Low production rate Low temperature Rapid reconstitution Better quality compared with drum drying

REFERENCES

1 B. Luh and J.G. Wood, Commercial Vegetable Processing, AVI Publishing Company, Westport, Conn., U.S.A., 1975.
2 M. Torrey, Dehydration of fruits and vegetables, Food Technology Review No. 13, Noyes Data Corporation, New Jersey, U.S.A., 1974.
3 L.P. Hanson, Commercial Processing of Fruits, Food Technology Review No. 30, Noyes Data Corporation, New Jersey, U.S.A., 1976.
4 S.D. Holdsworth, Dehydration in A.H. Johnson and M.S. Pederson (Eds) Encyclopedia of Food Technology, Vol. 2, p.282, AVI Publishing Company, Westport, Conn., U.S.A., 1974.
5 S.D. Holdsworth, J.Fd. Technol., 6(1971) 331-370.
6 W. Wolf and W.E.L. Spiess, Ind. Obst. und Gemuse, 61(14), 1976, 371-374.
7 G.R. Setty, K.B.S. Reddy, S. Saroja, K.C. Chikkappaji and A.M. Nanjundaswamy, Indian Food Packer, 32(1), 1978, 115-125.
8 N.D. Raev, S.S. Tanchev and R.C. Sharma, J.Fd. Proc. and Presv. 8(1), 1984, 15-29.
9 S.S. Tanchev, R.C. Sharma, Y. Vishansky, A. Mikhailova and I. Boyadzhiev, Proc. 6th Int. Cong. Fd. Sci. & Technol. 2, 1983, p.121.124.
10 T.L. Adambounou, F. Castaigne and J.C. Dillon, Sciences des Aliments, 3(4), 1983, 551-567.
11 R. Andreotti, M. Tomasicchio and L. Macchiavelli, Industria Conserve, 58(2), 1983, 90-95.
12 C.R. Lerici, G. Pinnavaia, M. Dalla Rosa and D. Mastrocola, Industria Alimentari, 22(203), 1983, 184-190.
13 M. Dalla Rosa, G. Pinnavaia and C.R. Lerici, Industria Conserve, 57(1), 1982, 3-7.
14 G.L. Mehta and M.C. Tomar, Indian Food Packer, 34(4), 1980, 12-15.
15 G.L. Mehta and M.C. Tomar, Indian Food Packer, 34(4), 1980, 8-11.
16 J.H. Moy and H. Spielmann, Research Bulletin No. 183, Hawaii Agricultural Experiment Station, 1981.
17 J. Hawkes and J.M. Fink, F.Fd. Proc. and Presv., 2(4), 1978, 265-284.
18 J.H. Moy, N.B.H. Lau and A.M. Dollar, J.Df. Proc. and Presv. 2(2), 1978, 131-135.
19 M.S. Ramamurthy, D.R. Bongirwar and C. Bandyopadhyay, Indian Food Packer, 32(11), 1978, 108-111.
20 S.S. Teaotia, G.L. Mehta, M.C. Tomer and R.C. Garg, Indian Food Packer, 30(6), 1976, 15-19.
21 C.R. Lerici, M. Pepe and G. Pinnavaia, Industria Conserve, 52(2), 1977, 125-129.
22 B.V. Zozulevich and Z.Y. Chumak, Konserv. I. Ovosh, Naya Prom. 4, 1974, 42-43.
23 J.D. Ponting, Process Biochemistry, 8(12), 1973, 18-20.
24 B.L. Flaumenbaum, Izrest. Vyss. Uchem. Zav. Pish. Tekh., 3, 1972, 187-188.
25 T.H. Jackson and B.B. Mohamed, Sudan J.Fd.Sci. & Tech., 3, 1971, 18-23.
26 M.A. Grishin, Izvest. Vyss. Ucheb. Zav. Pish. Tech., 4, 1970, 146-148.
27 B.V. Zozulevich and E.N. Dyachenko, Konserv. I. Ovosh. Naya Prom. 7, 1969, 32-34.
28 K.S. Jayaraman, V.K. Gopinathan, P. Pitchamuthu and P.K. Vijayarag-Lavan, J.Fd. Technol., 17(6), 1982, 669-678.

29 J.H. Moy and H. Spielman, Research Bulletin No. 183, Hawaii Agricultural Experiment Station, 1981.
30 J.H. Moy and H. Spielman, Food Technology, 34(2), 1980, 32, 34, 36-38.
31 T. Moreira, Technical Report, University of Dijon, France, 1976.
32 E. Maltini, Annali dell'Instituto Sperimentale per la Valorizzazione Technologica dei Prodotti Agricoli, 4, 1973, 89-107.
33 L. Calcagno, Industrie Alimentari, 12(10), 1973, 123-126.
34 L. Calcagno, Industrie Alimentari, 11(2), 1972, 65-68.
35 J.H. Moy, J.Fd.Sci., 36(6), 1971, 906-910.
36 G. Aceto, Freddo, 24(2), 1970, 3-9.
37 J. Tomas, F. Pinaga, B. Lafuente and E. Primo, Revista de Agroquinica y Tecnologia de Alimentos 10(1), 1970, 96-104.
38 M. Degoix, Rev. Pratique du Froid, 23(287) 1972, 41-43.
39 J. Tomas, F. Pinaga, B. Lafuente and E. Primo, Revista de Agroquinica y Tecnologia de Alimentos 9(3), 1969, 406-14.
40 J.P. Hayward, Food Industries of South Africa, 22(3), 1969, 45 and 47.
41 G. Aceto, Industrie Alimentari, 8(8), 1969, 86-94.
42 S.K. Chandrasekaran and C.J. King, Chem. Eng. Progress Symp. Series, 67(108), 1971, 122-130.
43 J.H. Moy, H.I. Oh and V. Gavino, Proc. IV International Congress Fd. Sci. & Technol., Vol. II(2), 1974, 36-39, Madrid.
44 A.R. Rahman, U.S. Patent 3, 984, 577, October 5, 1976.
45 V.A. Voskoboinikov, Z.A. Kats and A.I. Manuiko, Trudy. Vses. Nauch. Inst. Kons. i ovosh. Prom., No. 24, 1976, 37-43.
46 U.M. Khusainov, Fruit and Grape Dehydration using Accumulated Solar Energy, publ. Moscow, USSR, 1984.
47 S.K. Kalra and K.C. Bhardwaj, J.Fd.Sci. & Technol., India, 18(1), 1981, 23-26.
48 B. Kilkis, in 'Energy Conservation and use of renewable energies in the Bio-industries', 1981, pp. 307-333.
49 H.R. Bolin, C.C. Huxsoll and D.K. Salunkhe, Confructa, 25(314), 1980, 147-160.
50 B.A. Mohmood, W.H. Shah, F.H. Shah, in 'National Seminar on Dehydration, Preservation of Vegetables and Fruit', 1980, pp. 18-24.
51 W.M. Miller, in 'Changing Energy Use Futures', 1979, pp. 1893-1900.
52 R.L. Price, C.D. Sands, D.L. Larson, E.A. Mielke, G.R. Dutt in 'Changing Energy Use Futures', 1979, pp. 1910-1912.
53 S.C. Gupta, in 'Sun. Mankind's Future Source of Energy', Vol.III, 1978, pp. 2080-2085.
54 R.E. Berry, C.J. Wagner and R.L. Coleman in 'Changing Energy Use Futures', 1979, pp. 1871-1877.
55 T.M. Maksudov, G.G. Umarov, S. Artikov, Lebensmittel-Industrie 26(5), 1979, 209-210.
56 C.J. Wagner, W.L. Bryan, R.E. Berry, Proc. Florida State Horticultural Society, 90, 1977, 158-161.
57 T. Lucae, Z. Lebensm. und Verfahr, 29(7), 1978, 253-257.
58 J.F. Thompson, M.S. Chinnan, M.W. Miller and G.D. Knutson, J. Food Process. Eng. 4(3), 1981, 155-169.
59 L.A. Kabanov, Konserv. i Ovosh. Prom. No. 2., 1979, 27-29.
60 W.W. Mao, Doctoral Disertation, Cornell University, Ithaca, New York, U.S.A., 1975.

